TUFTING LEGACIES

Cobble Brothers to Card-Monroe:
The Story of the Men Who Revolutionized the Carpet Industry

Robert J. Tamasy

iUniverse, Inc.
New York   Bloomington

iUniverse books may be ordered through booksellers or by contacting:

iUniverse
1663 Liberty Drive
Bloomington, IN 47403
www.iuniverse.com
1-800-Authors (1-800-288-4677)

ISBN: 978-1-4502-5892-0 (pbk)
ISBN: 978-1-4502-5893-7 (cloth)
ISBN: 978-1-4502-5894-4 (ebk)

Printed in the United States of America

iUniverse rev. date: 10/13/2010

Dedicated to the hundreds of men and women who contributed in many ways to the tufting process and development of technology for the modern-day tufting machine.

Contents

Preface

Strolling through a retail store, you might pause to examine an item that catches your eye – a coat, for instance, an appliance, or a pair of shoes. How often does the thought come to your mind, "I wonder how they make this?"

If you're like most people, the question does not come up very often. After all, being able to enjoy a finished product doesn't require any understanding of the manufacturing techniques and processes utilized to produce it.

Because technological developments in our 21st century world come at us so rapidly, we tend to accept them and before long take them for granted. Understandable, perhaps, but still unfortunate because we fail to pay homage to the sweat of the brow, the determination of the spirit, and the ingenuity of the mind that all were so critical to the creation of goods that become central and commonplace in our daily lives.

Carpet is a classic example. Think of the last time you walked barefoot or in stocking feet across a plush, cut-pile carpet, or stepped into a room and admired the subtle tones and patterns of a cut-loop "sculptured shag" carpet. Did you take time to consider the trailblazers whose imagination, initiative and inventiveness helped make possible such versatile and diverse floor coverings? Except for those directly involved in the carpet business, few people ever do. We all, however, appreciate the end result.

A number of books have been written to document the history of the carpet industry from an overall perspective, and some have served as helpful resources in preparing to write this book. However, they concentrate primarily on the companies and enterprising individuals that helped to advance the manufacturing and retail aspects of the business. Comparatively little focus has been given to the innovators who one small, incremental step at a time assembled, adapted, modified and refined tufting machines into the fast, efficient instruments of mass production that they have become.

The unlikely journey from the first crude, hand-tufted bedspread at the dawn of the 20th century to the multicolored, patterned, varied textured pile carpets of today almost seems like the figment of someone's fanciful imagination. One might be tempted to dismiss it, concluding "you can't get there from here" – except it's true. Starting in the 1930s, men bearing the last names of Cobble and Card – with the assistance and support of many individuals along the way – slowly and systematically pursued dreams that first seemed unreachable, then improbable, until they started to seem possible, then probable, and finally, attainable. To these gifted and ingenious individuals we owe a tremendous debt every time we set foot on carpet in a home, an office, or a store.

Beyond that, their curiosity, vision and innovation have had an immeasurable impact on a region that has been greatly enriched by the carpet industry, as well as socioeconomics and everyday lifestyles across the United States and, increasingly, around the world.

This book is, in part, an attempt to direct much-deserved attention to the lives and unique contributions of six special individuals: Albert and Joe Cobble, who laid the groundwork; Lewis and Roy Card, who experimented, guided and perfected the work; and finally, Lewis Card Jr. and Charles Monroe, son and son-in-law, respectively, who are among those continuing the work today.

Without question, hundreds of people have made their own notable contributions to the development of the carpet industry over the past 50 years or so. Some assisted in exploring and developing the intricate technology that helped to make tufting machines the marvels they are now. Others applied the capabilities of that machinery to extend the limits of carpet design and quality. Some will be acknowledged in the chapters that follow, although even attempting to mention every individual would be impossible.

These pages, however, are reserved primarily for celebrating the special place the Cobble and Card families hold in the carpet industry's colorful evolution.

A BASF commercial on TV declares, "We don't make the carpet. We make the carpet better," referring to its chemical research to enhance the quality, color and durability of modern-day carpet fibers. Card-Monroe Corp. (CMC), a leader in the tufting machine industry today, could adopt a similar motto: "We don't make the carpet. We make the *machines* that make the carpet – and we make them better."

But there would be no CMC if first there had not been a Cobble Brothers Machinery Company, or Super Tufter, Singer-Cobble, Roy T. Card Company, Southern Machine, or Tuftco. In reading this book you hopefully will gain a much greater appreciation for the mechanical processes and technology utilized by the machines that every day churn out miles of carpet to adorn our private dwellings, centers of commerce and public facilities. But even more important, you will also gain a sense of the singular achievements of a handful of men – at once inventors, innovators and leaders of industry – whose minds, skills and personal resolve made these machines possible.

Robert J. Tamasy
Chattanooga, TN
2010

Acknowledgements

As with any book, this story about the development of the tufting machine and the people who led the way would not have been possible without the help and contributions of many people. We would like to express our appreciation to:

- Marcelle White, Executive Secretary of the Whitfield Murray Historical Society, for vintage photos and her perspective on the early days of tufting.
- Landmarks of DeKalb County, Alabama for photos of Fort Payne, Alabama in the 1930s and 1940s.
- The Carpet and Rug Institute in Dalton, Georgia for photos and books that provided valuable background on the history of the carpet industry.
- The Chattanooga-Hamilton County Bicentennial Library for vintage photos of Cobble Brothers Machinery Company and Singer-Cobble.
- The Shaw Industries Carpet Research and Development Center museum, for photographic access to historic tufting equipment.
- Med Dement, whose photos make up the cover of this book.
- Lois Hoffmann for her diligent and detailed proofreading and editing of this manuscript.
- Mitzi Young for her help in coordinating the numerous meetings required throughout the process of this book.
- The countless men and women who contributed to the development and refinement of the tufting process in many ways. It is impossible to name them all, but they each played an

important part in bringing the craft of carpet tufting to where it is today and rightfully deserve to be a part of this story.

- The late Harold North, who passed away before this book could be published. He made many invaluable contributions in expanding the market for tufting machines worldwide.
- Buddy Cobble, for providing photographs of his father, Bud Cobble, and grandfather, Albert Cobble.
- Gene Duff, for providing information documenting history of the carpet industry over the past 50 years.
- Cessna Decosimo, sculptor of the busts of Joe Cobble, Lewis Card and Roy Card, which are displayed in the main lobby at Card-Monroe Corp.
- Barry Aslinger, for photo of Lewis and Roy Card being inducted into University of Tennessee-Chattanooga Entrepreneurship Hall of Fame.
- Ken Johnson, who encouraged us and had helpful advice for launching this book project.
- John Potts and the team at iUniverse, who patiently waited for this manuscript to be completed and helped to guide the process into print.

1   Spreading a Tufting Revolution

Carpet.

When you hear that word, what ideas or images immediately come to mind? At one time carpet served as a symbol of prestige and exclusivity. Only the wealthiest could afford soft floor coverings that were produced by the tedious, time-consuming, costly craft of weaving. Today, however, thanks to the tufting process that revolutionized the industry, carpet has become as integral to American culture as football in the fall, McDonald's, Mount Rushmore, and 4th of July fireworks.

We find carpet everywhere: Luxurious hotels and sprawling estates. Beach bungalows and suburban subdivisions. Lavish apartments and assisted living quarters. Schools, office buildings, restaurants and shopping malls. Economy hotels. Mobile homes. Even cars, trucks, vans and SUVs. We regard carpet as a certainty of everyday life, much like taxes and the changing seasons.

Signs along I-75 about 30 minutes from Chattanooga, Tennessee point travelers to an amazing carpet retailing hub: Exit 328, just south of Dalton, Georgia. There they can find Carpets of Dalton, Georgia Carpet Industries, Beckler's Carpet, and dozens of other shops large and small, featuring carpet in limitless varieties, all situated strategically along what is often called "Carpet Alley" (Dug Gap Road).

Visitors see an astounding assortment of carpet styles – cut pile, loop pile, cut-loop, high-low loop, shags of various lengths and bulky berbers. Samples are presented in a spectrum of reds, blues, beiges, browns, tans, whites, grays, greens, golds, yellows, purples, pinks and oranges. They range from brilliant, bright and bold shades to subdued earth tones.

Patterns and designs come in countless variations – diamonds, squares, triangles, circles, zig-zags, swirls, stripes and other shapes. Some feature cartoon and fantasy characters, animals, or team logos. There is even a style called "shaggy raggy," consisting of flat and wide pieces of multi-colored linen fabric tufted for use in children's and adolescents' bedrooms and play areas.

But there's still more: Possible uses for tufted carpet have stretched the horizons of imagination. Many of us can remember the days of rock hard, abrasive Astroturf that was developed for indoor playing fields. Today the tufting process is being implemented to produce a much more grass-like artificial turf for football, baseball and soccer fields, indoor/outdoor environments, play areas, home lawns, landscaping, golfing ranges, and even streetscapes.

Few Carpets – or Even Rugs

Tufted floor coverings have become so prevalent, so readily accessible that one might easily conclude the right to carpet ownership must have been incorporated somewhere in the Bill of Rights, alongside freedom of the press, freedom of worship, and the rights to vote, assemble, and bear arms. But amazingly, less than two generations ago, this was hardly the case. In fact, prior to the 1950s, most homeowners considered themselves fortunate to possess even a lonely area rug, or perhaps a few throw or "scatter" rugs. Hardwood floors were not an interior design option; they were standard features in most dwellings.

Per capita carpet consumption had actually declined during the first half of the 20th century, and trade journals and carpet mill executives openly expressed doubts about the industry's future. In 1951, the overall volume of the tufting industry – which included bedspreads, rugs and carpets – amounted to only $113 million. By itself, carpet accounted for only $19 million of that total.

Then during the '50s, not even six full decades ago, a quiet but dramatic manufacturing revolution began. "It was as if someone had opened a magic trunk," the Carpet and Rug Institute states on its web site. "Out of that trunk came manmade fibers, new spinning techniques, new dye equipment, printing processes, tufting equipment, and backing for different end uses."

This manufacturing and technological revolution slowly built momentum until tufted carpet burst upon the American scene full force in the 1970s and '80s, gaining a grip on home decor at all levels of society that it maintains to this day.

Gradually, the fervor for tufted carpet also began to take hold beyond American shores, reversing the early colonists' journey by crossing the Atlantic Ocean to Great Britain and Europe. Today, carpet is regarded as "the floor covering of choice" in all developed countries, including Latin America,

Australia, the former Soviet Union, and extending even into China and parts of the Middle East.

The Carpet and Rug Institute, which has its headquarters in Dalton, reports nearly 92 percent of all carpet is now produced by the tufting process. A classic example of mass production, approximately 1.5 billion square yards of tufted carpet are manufactured annually (compared to only 97 *million* square yards in 1950). Putting this statistic into visual terms, one billion yards of carpet would comprise a roll 12 feet wide circling the earth at the equator nearly six times, or stretching two-fifths of the distance to the moon.

Total carpet industry sales today exceed $14 billion, and more than 80 percent of the total U.S. carpet output comes from mills in North Georgia. Perhaps the most dramatic example of the industry's transformation is Dalton's Shaw Industries, now owned by Berkshire Hathaway, Inc. Shaw grew from a $300,000 finishing company in 1958 into the largest carpet manufacturing company in the world, with annual sales in excess of $5 billion.

In actuality, this relatively recent phenomenon could never have taken place without the vision, perseverance, innovation and genius of two remarkable families from Chattanooga and Southeast Tennessee – the Cobbles and the Cards. More specifically, it involves the compelling tale of three unique pairs of individuals who have collaborated as brothers, in a literal or figurative sense – Joe and Albert Cobble, Lewis and Roy Card, and Lewis Card, Jr. and Charles Monroe.

But before starting to recount the immense impact these industrial trailblazers and pioneers have had on the carpet industry through their development and refinement of tufting machinery, it's helpful to take a brief journey back in time to the turn of the century – the *20ᵗʰ century* – to establish a context for their accomplishments.

Dating Back to Ancient Egypt

According to historians, carpet is hardly a modern conception. Some experts on antiquities have traced the earliest carpets to the ancient Egyptians, the Babylonians and later the Roman Empire. From there the fascination among the privileged with these special floor coverings spread to Western Europe, where weavers combined skill and artistry on elaborate looms.

Even in the United States, carpet dates back to the Colonial era when industrialists from England and Europe imported the product for the first American aristocrats. For more than 150 years, "manufactories" in the Northeast produced intricate, labor-intensive woven floor coverings that would be showcased exclusively in homes of the rich and genteel. To create luxurious pile, each small tuft was made by a very slow, mechanized process, placing a high premium on time, expertise, and expense.

Carpet for the common people, therefore, seemed as probable as a colonist returning to his or her European homeland by navigating the Atlantic in a rowboat. As historian Anthony N. Landreau observed, "Colonial Americans busy carving out a new continent had little time to enjoy such amenities as rugs on the floor."

Around 1892, however, a seemingly innocuous introduction took place. This proved to be the catalyst for an incredible chain of events, starting with a simple wooden spinning wheel and culminating in the huge, 2,500-pound, computer-controlled tufting machines that have made the carpet industry what it is today.

In that year, a 12-year-old girl named Catherine Evans, born and raised in a quiet hamlet not far from Dalton, visited a cousin in nearby Trion. There she took note of a family heirloom, a bedspread that had been fashioned by hand in years predating the Civil War. "I admired it so much," Catherine would recollect many years later. She immediately determined that "when I grow older, I'm going to make me one."

British and French immigrants had brought this "tufting" craft to Colonial America, with women in New England utilizing it for bedspreads and other home decorations. The art of tufting slowly advanced into the South in the 18th and 19th centuries, but had declined and virtually disappeared prior to the War Between the States. That is, until young Catherine happened across the remnant that caught her eye.

Recapturing a Lost Art

The distinctive tuft-forming effect on this well-preserved bedspread stirred her imagination. By 1895, hoping to replicate the spread's unique handiwork, young Catherine began to experiment with fabric and technique. Starting with unbleached cotton sheeting, she copied a quilt pattern using a stencil to create an artistic design on the sheet. Then she followed the pattern by hand with a needle and heavy, 12-ply yarn she had spun on the family's old spinning wheel.

After finishing the sewing portion, Catherine clipped between the stitches, forming yarn ends that extended above the sheeting so they could produce a puffy, "chenille" effect. She then washed the bedspread several times in boiling water, shrinking the fabric and thereby locking in the yarn tufts. Lastly, the bedspread was hung on an outdoor clothesline to bleach and dry in the sun and "bloom" in the gentle breeze. The finished look came to be known as "candlewick," since the flared and fluffed tufts resembled cotton wicks typically used in candles of the day.

At the time, Catherine never dreamed that her captivation with the tufting technique could become the first step in an eventual flooring revolution. (Many years later she would witness firsthand the culmination of her creativity, watching carpet roll off huge tufting machines.) Her only concern initially had been persevering through the meticulous process she devised to create a solitary bedspread. Perhaps based on past behavior, her mother had expressed doubt that Catherine possessed the stick-to-itiveness to complete even that initial spread.

Disproving her mother's prediction, the teenager did finish the tufted bedspread. In fact, Marcelle White, executive secretary of the Whitfield Murray Historical Society, has offered the view that the craftsmanship of the colorful fabric was actually far superior to the pre-Civil War spread that had served as Catherine's inspiration.

Embracing this hobby almost as a "calling," she tenaciously proceeded to make a second tufted bedspread and presented it to her brother and his bride as a wedding gift in 1896. Understandably, such an unusual and creative gift could not escape notice. Before long Catherine started receiving orders to produce more. The first bedspread she made for sale fetched for her the phenomenal sum of $2.50, which she almost apologetically itemized as $1.25 for materials and $1.25 for her labor.

Creating each bedspread, however, required the same painstaking, laborious process. While she functioned as the original "tufting machine," her work was slow and inefficient. The young woman managed to develop some shortcuts to speed up the process somewhat, but before long she was uttering a plaintive, "Help!" Out of that desperation an industry was about to be born.

Passing on Her Secrets

Catherine began teaching the art of making the bedspread to women living in her area, unselfishly eager to pass along the secrets of her craft. Locals commonly referred to the technique as "turfting" or simply, "turfin'." She would stamp patterns on the muslin sheeting – initially using can lids or pie tins and grease from meat skins, and later dye-colored wax markers – and then distribute the material to local women who would employ their new skills to finish the work.

Catherine Evans Whitener is widely credited for rediscovering and
advancing the craft of tufting, resulting in a cottage industry of
handmade bedspreads in northwest Georgia in the early 1900s.
(Photo courtesy of Whitfield Murray Historical Society)

By the time she married W.L. Whitener in 1922, Catherine Evans
Whitener had matured into an accomplished businesswoman, fostering a
regional entrepreneurial movement enfolding the efforts of many dozens of
households that diligently fashioned bedspreads out of the yarn, sheeting and
patterns she brought into their mountain homes.

A major breakthrough occurred in the 1920s when one enterprising
woman was bold enough to try selling the bedspreads to large department
stores in the North. When the John Wanamaker retail store in Philadelphia
agreed to buy 15 of the spreads for $98.15, the handicraft started to build
momentum. By the 1930s, demand for the decorative spreads had transformed
the one-time avocation into a true Northwest Georgia cottage industry, with
entire families engaged in the work.

Spinning wheels were used to produce yarn for the early tufted
bedspreads made in northwest Georgia in the early 1900s.
(Photo courtesy of the Carpet and Rug Institute)

With the United States wallowing in the throes of the Great Depression,
husbands and fathers in the Dalton area found themselves hard-pressed to meet
their families' needs solely through farming. To supplement their incomes,
they would return from the fields at night and work alongside their wives and
children to complete the bedspreads. As Thomas M. Deaton comments in his
book, *Bedspreads to Broadloom: The Story of the Tufted Carpet Industry,* "the
emerging bedspread industry was a lifesaver to the 'dirt-poor' farmer."

Haulers – trucks, wagons, even mule carts driven by men – would
make rounds of "spread houses" where the bedroom novelties were being
made, loading up the finished goods and carrying them to nearby Trion and
Summerville, to markets in other Southern states, and eventually northward
to adorn more prosperous households in the Northeast.

Brightly colored bedspreads also were displayed and marketed locally,
hung on clotheslines and sold from roadside stands that could be spotted
by travelers on U.S. Highway 41, between Dalton and Cartersville, which
became known as "Bedspread Boulevard." (The heavily traversed highway
corridor, used by people from as far north as Michigan's upper peninsula and
as far south as Florida's lowest tip, also acquired the nicknames of "Bedspread
Alley," "Chenille Alley" and "Peacock Alley" – the latter referring to the
extremely popular design of two peacocks facing each other.) This appealing
concentration of handcrafts also earned for Dalton, formerly known mostly for
its cotton mills, the designation of "the Bedspread Capital of the World."

Tufted bedspreads like these were displayed along U.S. Highway 41 near Dalton, Georgia beginning in the 1930s to entice travelers. The route was known as "Peacock Alley" for the colorful designs. (Photo courtesy of the Carpet and Rug Institute)

Propelled By Shifting Paradigms

At this point, the primitive tufted spreads remained light-years away from the tufted carpet that would become such a mainstay of the latter 20th and early 21st centuries. However, several factors came into play that drastically shifted the bedspread industry's paradigms and began to transform it into a significant player in the world of manufacturing.

The first factor was mechanization. It became evident that the hand-stitched method was far too snail-paced to keep up with the escalating demand. As men increasingly became involved and recognized the retail potential for bedspreads, they began searching for faster, mechanized solutions to the problem. The result was the first tufting machine.

Historians believe a crude tufting machine may have been fashioned as early as 1922, with inconsequential attempts dating back even to the late 1800s. But no one knows for certain who invented the first functional tufting machine. Several individuals seemed to have arrived at similar conclusions almost simultaneously – men like August Carter of Chattanooga and Ernest Moench of Nashville. Apparently they had experienced common "brainstorms," envisioning the conversion of an industrial sewing machine, such as the Singer 3115 model, by adapting it to hold multiple needles, hooks and knives to form a continuous row of cut pile stitching.

By the early 1930s, a number of men felt deserving of the right to be recognized as the true inventor of the tufting machine. To illustrate this confusion, R.E. Hamilton, one-time head of the Tufted Textile Manufacturers Association, quipped that if you were to invite the man who invented the machine to meet you on the steps of the courthouse in Dalton, it would be impossible to go up the steps – people claiming that honor would be standing elbow to elbow with no wiggle room.

One of the most popular choices was Glenn Looper of the Looper Foundries in Dalton in 1936. He reportedly once told a journalist that his father-in-law had complained that despite the Kenner-Rauschenberg Bedspread Company's high volume, it still was failing to realize a profit. "Can't you build us a machine to do this tufting work?" he had pleaded.

Looper is credited for creating a single-needle tufting machine that inserted large yarn into a fabric and cut the tufts with a scissors-like mechanism. The hooks on tufting machines utilized to catch or pick up the yarn after the fabric is penetrated by the needle – commonly known as "loopers" – are said by some to have been named after him, although the term also describes their function. However, his claim as *the* inventor is hardly indisputable; industry historians acknowledge that others of the period also merit consideration.

From Spread Houses to Factories

Ultimately, the importance from an industrial perspective rested not on the "who" but the "what": With the introduction of mechanization, even though only a single needle was in operation at first, tufting had taken a significant step toward mass production. This enabled many established spread houses to expand into bonafide factories.

A second impetus toward mechanization was the National Industrial Recovery Act, enacted by President Franklin D. Roosevelt in 1933 to provide relief for underpaid, overworked laborers in various industries. In particular, under the act's provisions, "hand tufters" who had been receiving 5-10 cents per hour were mandated a minimum hourly wage of 32.5 cents. Maximum daily work hours also were reduced substantially.

As employers fretted over shrinking profit margins, they responded by seeking to develop machinery that could reduce production costs and increase output. At this point, geographic proximity played a significant role: The tufting industry would sink deep roots into the "Tri-State" area of northwest Georgia, northern Alabama and southeastern Tennessee as it expanded into the production of other textiles, such as bathrobes, scatter rugs, bathmats and, ultimately, carpet.

In the early 1900s, Chattanooga – situated just about 30 miles north of Dalton – was regarded as the most industrialized city of its size in the United States, boasting nine furnaces, 17 foundries and numerous machine shops, along with textile mills. When bedspread companies began experimenting with machinery, they often turned to machinists in Chattanooga for parts and expertise in building their machines.

Fort Payne, Alabama, located about 50 miles southwest of Chattanooga, also served as home for a number of hosiery mills. Mill owners and mechanics in both industrial centers – including Albert and Joe Cobble in Chattanooga – were attracted to the emerging tufting industry. The tufters and machinists forged working partnerships that would begin to propel the industry toward horizons these ambitious innovators never could have envisioned.

In *Bedspreads to Broadloom,* Deaton observes, "The story of the carpet industry is not merely one of square yards produced, but it is also the story of the people who worked, dreamed, and created the companies, the machines, and the carpet." Although many individuals played notable roles in the revolution – and evolution – of the industry, particularly in its early years, soon the names of Cobble and Card would rise to special prominence.

As we begin to consider the unique and invaluable contributions the men of the Cobble and Card families made in helping to transform carpet into what it has become for modern living today, it's important to take a side trip down "Memory Lane," including a stop at Fort Payne and proceeding on to Chattanooga, which in the early 1900s were a relatively short train ride apart.

2

FERTILE MINDS IN FORT PAYNE

Having the brief historical overview of carpet, the saga of Catherine Evans Whitener and the birth of tufting as background, we can now begin, to borrow the familiar phrase from the late radio commentator Paul Harvey, "The rest of the story": How a young woman's simple notion for colorfully adorning a solitary bedspread became the impetus for a convergence of events that culminated in the development of an increasingly sophisticated process for making luxurious, yet affordable floor coverings for people from virtually every walk of life.

The tufting industry's earliest roots will always be traced to Northwest Georgia and Dalton, where hardy, independent settlers had forged proud, self-sufficient lifestyles. The resourcefulness which enabled them to live off the land was redirected to initiate unique industries that ultimately would have a dramatic and lasting impact on American society and culture from coast to coast.

Over the past half century, Dalton has assumed a new identity, morphing from its one-time reputation as "Bedspread Capital of the World" into "Carpet Capital of the World." During that same span, Chattanooga, Tennessee has risen into preeminence as the hub of the tufting machine industry, becoming the focal point for the steadily advancing technology that has created the modern-day mechanical marvels that produce miles upon miles of carpet every day.

However, to be complete, any discussion of the tufting industry's early days should include some mention of Fort Payne, Alabama. Even though there were never any tufting enterprises in this Southern community, it still figured indirectly in the genesis of that stellar example of American ingenuity, enterprise and perseverance. Perhaps best-known today for being the hometown of "Alabama" – the legendary country-western vocal group –

Fort Payne for decades also has been recognized for its contributions to the world of American textiles.

Situated approximately 50 miles southwest of Chattanooga, DeKalb County (originally known as Will's County) was established in 1836. Fort Payne, the county seat, became incorporated in 1889, taking its name from Capt. John G. Payne, commander of the garrison that housed Cherokee Indians before their forced march in 1838 to Oklahoma along what is now widely known as "The Trail of Tears."

From 'Boom' to Bust, and Back to Boom

Rich coal and iron ore deposits discovered in 1885 temporarily transformed Fort Payne into a boom town. However, those mineral reserves became depleted in 1893, turning boom into bust, and mining companies migrated southward to richer veins in Birmingham's Red Mountain. Fort Payne's local citizens were left behind to eke hardscrabble livelihoods from the land, but this changed markedly when the town's first hosiery mill opened some 14 years later.

In 1907, the Florence Knitting Company acquired the former site of the Alabama Builders Hardware Manufacturing Company, also known as the Foster Building. (Constructed by the Fort Payne Coal and Iron Company, the hardware plant had been abandoned after mining interests departed from the area.) On Oct. 16, 1907, a deep-toned whistle signaled the opening of the hosiery mill at 6:30 a.m. About 30 knitting machines and 12 finishing machines soon burst into motion to knit socks on the building's first floor, launching a local industry that has continued into the 21st century.

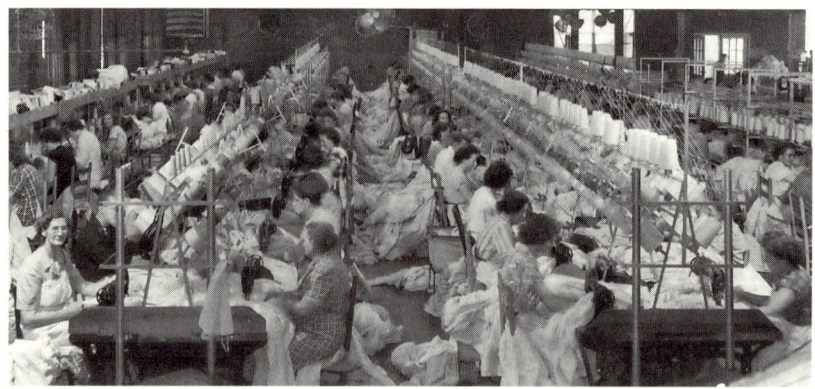

Women made up most of the workforce making chenille
or tufted bedspreads in the 1930s. (Photo courtesy of
Chattanooga-Hamilton County Bicentennial Library)

Almost immediately, the mill became the town's No. 1 employer. Most of the first employees were women and girls who operated the machines, earning meager sums of 10 to 17 cents per hour. Low as it was, that rate was better than the 50-cent daily compensation many men earned by working on farms. And with no other industrial jobs available in Fort Payne, a job at the mill was considered a tremendous opportunity. Men were hired to perform the duties of foremen, superintendents and "fixers" – mechanics who responded quickly to repair machines when they broke down and replace worn-out parts. Cultivated skills like these eventually would factor into the development of the tufting industry in nearby cities.

The Davis Mill or "the big mill," where hosiery was made, was the largest employer in Fort Payne during the early 1900s. (Photo courtesy of Landmarks of DeKalb County, Alabama)

In 1915, the mill became known as the W.B. Davis Hosiery Mill when its namesake, W.B. Davis, elected to increase his 10 percent ownership in the mill, exchanging his shares in Chattanooga's United Hosiery Mills to become the majority stockholder in the thriving Fort Payne factory.

Over time, Davis's hosiery mill acquired the nickname of "the big mill" as other smaller mills were established near the downtown area to accommodate the burgeoning local industry. The mill's payroll served as the principal source of cash flow in the community for many years.

Hosiery mills remained at the heart of the town for decades. In fact, similar to its counterpart to the east – Dalton – Fort Payne earned a global reputation through its own industrial expertise, becoming known as "the Sock Capital of the World." It embraces that title to this day, ranking as the largest single location for hosiery manufacturing in America even though, as has become true of most textile products, socks are now made primarily outside the U.S. borders.

Moving in Search of Opportunity

It was this growing industry that attracted members of the Cobble and Card families to Fort Payne. The Cobbles were a hard-working clan that traced their ancestry to Germany. Joseph Henry Cobble and his family – including sons Henry, Horace, Albert, Joe and Roy – had established their home in Sweetwater, Tennessee, about 50 miles northeast of Chattanooga. Eventually the sons moved to Chattanooga, and several continued on to Fort Payne where they became involved in the hosiery industry.

This Cobble family photo dates back to the late 1800s.

The Cards have traced their American family roots back to approximately 1640, when John Card sailed across the Atlantic Ocean from Devonshire, England. Around 1830, Captain Edward Shannon Card, Jr. and his family settled in Soddy-Daisy, Tennessee, a community about 20 miles northeast of Chattanooga. There one of his sons, Charles Parker Card, whom family historians have described as a man of strength and vision with a fierce determination to shape his own destiny, emerged as a patriarch for future generations of Cards.

When the Great Depression struck in 1929, its harsh economic realities left no room for sentimentality – "home" became wherever work could be found, and families readily relocated in the pursuit of employment opportunities. So despite nearly 100 years of family history in the Soddy-Daisy area, Russell and Anna Mae Cobble Card and their brood of two boys and three daughters moved from Daisy, Tennessee into the simple, yet ambitious environment of Fort Payne in October 1930.

A scene from a Card family gathering in the late 1800s..

Russell had been working as a fixer in a hosiery mill in Chattanooga, but was always seeking ways of improving himself vocationally and providing more substantially for the needs of his young family. One of Anna Mae's brothers, Henry Cobble, was running a hosiery mill in Fort Payne – having been recruited from Chattanooga by W.B. Davis – and had offered Russell a full-time job as a fixer.

The Fort Payne of 1930 boasted a citizenry of about 5,000 people. While most of its people worked in the hosiery mills, a Coca-Cola bottling plant served as an employer of secondary standing. Agriculture (particularly cotton) remained important to the area, and during the Depression, the Civilian Conservation Corps camp near Fort Payne provided additional modest employment.

Lewis Card, the eldest sibling, was 10 years old when the Cards moved there, while Roy, the youngest, had just been born the preceding August. In between were three sisters, Ruth, Betty June ("Bett"), and Selma. To supplement the family's income, Anna Mae worked as a seamstress, made all of the children's clothes, and took in ironing for other people. It was hardly a time of abundance for the Card family, but neither was it perceived as a time of destitution.

Plenty to Eat and Money for 'The Show'

"Daddy always had a job, even during the Depression. We did not have a lot of things, but then we didn't know anybody that did," recalled Ruth Card Pickett.

"We always had a cow, a garden and plenty to eat," agreed Bett Card Bracken. "We never felt deprived. A lot of other people my age remember it as a time of suffering, but I don't. We had good parents and a very enjoyable life. Our mother's greatest concern was always to have money to pay for us kids to go to the show (one of the local movie houses)."

Those included the Strand and DeKalb theaters, located less than two blocks apart. The Strand was the older of the two, and the DeKalb was opened about six years after the Card family arrived in town. (Fort Payne's first motion picture theater, originally an opera house built in 1889, had been shut down before the Cards took up residence. It was reopened years later and became designated a national historical site.)

Typical showings included westerns and serialized movies like "Flash Gordon," and children often spent several hours each Saturday at the theater, watching the feature film two or three times before returning home. The Cards' only parental stipulation was, "Be home before dark."

Downtown Fort Payne, Alabama in the 1940s. (Photo courtesy of Landmarks of DeKalb County, Ala.)

"Fort Payne was such a small town then, a great place to grow up. You knew everybody in town and felt welcome everywhere. You could sleep with your windows and doors open and not have to worry. On Saturdays, in addition to going to the movies, we would pass the hours by walking uptown and back several times," according to Ruth.

For most of their years in Fort Payne (1930-1943), the Card family lived "down in the country," a few miles south of downtown, but they also spent some time "up in the country" (north of downtown).

"We count seven or eight houses that we lived in – a couple of them we lived in twice. Often it was our mother trying to find better quarters for us to call home, but sometimes I think she just liked to move!" quipped Bett.

Besides the movies, the Cards and their friends found a variety of other diversions to occupy their days – usually barefooted. "No one wore shoes during the summer," Roy noted. "Those were reserved for school." Added Lewis, "Sometimes those shoes didn't have soles in them, or the soles had big holes. That was just the way things were for 98 percent of us in Fort Payne.

We were all alike – we just thought that was the way it was, and how it was supposed to be."

Their leisure activities included going to the local roller-skating rink or roller-skating on a concrete slab nearby where a dry cleaning shop had burned down. There was a ball field several blocks away where the boys spent a lot of time. The siblings sometimes hit a ball back and forth on a dirt tennis court close to their home, and typically spent warm Sunday afternoons frolicking in the "blue hole" – an inlet that came off the Little River – under the observant eyes of their parents.

"We all learned to swim by going to the river. At first our father would watch us closely, but once we seemed to get the hang of it, he would just throw us in and say, 'Now swim!'" Ruth recalled.

When they felt adventurous, the children balanced along the narrow ledges at the bottoms of billboards, including one near the Card home. Bolder youngsters even tried walking across the tops of the billboards – but not the Cards, who were instructed to avoid such risky behavior. That warning, however, apparently did not keep the Cards from joining their friends in occasionally reenacting scenes from war movies they had seen. They would jump off the roof of their uncle Albert's garage, holding onto handkerchiefs and using them as "parachutes."

Stirring Fertile Imaginations

Most of all, the 1930s were a time for developing fertile imaginations. Years before anyone would experience the wonder of television, radio programs filled their evening hours. The Card children's minds were enthralled by the hilarity of "Amos and Andy," the adventures of "Sky King," music from "The Lucky Strike Hit Parade," and the suspense of "The Green Hornet" and "Inner Sanctum." With only sound effects to help in creating mental pictures, their young imaginations were constantly stimulated and nurtured.

During daytime hours they sometimes participated in a game called "Play like...," in which they would make believe ("play like") they were cowboys and Indians, truck drivers, or airplane pilots. One "play like" occasion was particularly memorable for the Card children, as we will see later in this chapter.

Unlike today, when games come out of a box, get downloaded onto a computer, or are inserted into a machine to be viewed on a monitor, entertainment for the Card children in the 1930s was a byproduct of their individual creativity. This likely contributed substantially to the ingenuity and innovation that Lewis and Roy would need to utilize later in their lives.

These unfettered imaginations were bolstered by an innate mechanical aptitude that most of the Cards shared.

"Lewis and Roy proved to be the most accomplished members of our family in terms of mechanical ability, but I think it was a trait they inherited, something in their genes," Ruth stated. "Our grandfather, Joseph H. Cobble, was a sharecropper, and his work on a farm required mechanical know-how. Our other grandfather, Lewis Audley Card, was a carpenter. Our uncles had all been in the hosiery business, working on the machines, as our father did.

"Even our mother, Anna Mae, had the ability to fix things whenever it was necessary. She was great at sewing. She could look at a dress in a store, then come home and make it almost exactly as she saw it. She had the same analytical, craftsmanship-oriented mind that later enabled our brothers to accomplish what they did with tufting machines.

"She liked to read to us and encouraged us to read, and was well-spoken, even though she had little formal education. She never let that be an impediment."

A Mind to Work

A strong work ethic was another Card family characteristic that helped to prepare Lewis and Roy for their careers. "Mama always said our job, as children, was going to school. But everyone in our family was a hard worker. Laziness was not a trait that we ever experienced or observed in our home," noted their sister, Selma Card McElhaney.

Anna Mae's own work ethic had been established early in life. At the age of 10, she had started working in a hosiery mill in Sweetwater. She was so small, according to family lore, that she had to stand on a box to run a knitting machine. Ultimately, she forged a lifelong relationship with hosiery mills, continuing to work in a mill in Chattanooga with her brother, Roy Cobble, until she was 80 years old.

During the Depression era, survival – not success – was the goal, and jobs were not taken for granted. When the Cards decided they needed to move back to Chattanooga in 1943, where Russell had gotten a job at a TNT plant that made munitions for the war effort, Bett was a senior at DeKalb High School in Fort Payne, at the foot of Lookout Mountain. That fact was regarded as little more than an inconvenience. "You had to move where the job was," she understood.

Being the oldest sibling, Lewis had started early in working at a number of jobs around Fort Payne to help supplement the family's income, including carrying newspapers and working at a filling station at night, before taking a job at the big mill.

A Singer industrial sewing machine, used to produce chenille rugs in the 1930s, on display at the Shaw Research & Development Center.

"Lewis was a hard worker for as long as I can remember," Ruth said. "There was one morning when it was so cold that his boots almost froze to his feet while he was delivering newspapers. Our mother had to drive him in the car to help him finish his route."

He enjoyed sports, but in high school Lewis's involvement with the football team had to be limited to the role of water boy due to a childhood accident he had suffered.

While visiting with the Cobble family in Sweetwater when he was about four years old, Lewis had been running to catch up with an uncle and one of his cousins when he stumbled. His right eye became punctured by a nail, and by the time they took him back to the house, he already had lost a lot of fluid from the eye.

"In those days we didn't have emergency rooms or easy access to hospitals, so my parents couldn't take him to the doctor until the next day," Ruth recalled. "By then there was no chance of saving his eye. I don't remember a lot about that incident, but I have often thought about what Lewis went through as a child. But he never let that disability hold him back or temper his determination to succeed at whatever he did."

Lewis wore a patch over his eye throughout his childhood, and upon graduation from high school received a very unusual congratulatory gift – a glass eye from his parents.

When he was about 16, he had also been assigned the task of overseeing his sisters and brother when their mother went to work at the mill full-time. "She put me in charge, but Lewis was really in charge of all of us," Ruth pointed out. "He wasn't too stern, but we knew we all had to mind him."

Although he loved his family, like many teen-aged boys Lewis was not inclined to show it, as sister Bett vividly remembered. "When I was about eight years old, I was playing a fairy in a school play. I even wore a tutu. Mother had to work, so she asked Lewis (then about 15 or 16) to take me to school. He walked in front of me all the way. It took me six or eight steps to match one of his – and I was going as fast as I could to keep up. If I had gotten into trouble, I know he would have been there to help me in a flash, but he wasn't about to be seen walking with his little sister."

Flirting With Disaster

Being 10 years apart in age, Lewis and Roy did not have a close relationship while they were growing up. While Lewis was rapidly maturing at home, Roy was progressing through infancy and his toddler stage. Only on rare occasions was little brother allowed to tag along with big brother. Lewis was already setting his mind for the realities of adulthood. For them, a close brotherly relationship would not develop until years in the future.

During his own childhood years, Roy seemed to flirt with danger as a moth flits around a flame. Ruth recalls he became extremely ill for about one week when he was 18 months old. Because access to advanced medical care was so limited, in those days the cure for an undetermined illness often consisted of simply riding it out and hoping for eventual recovery.

"My mother and I took turns caring for him," Ruth said. "We never really determined what had been wrong with him, but he was a very sick little boy for that week. To this day I tell people that Roy is such a good man because as his oldest sister, I virtually raised him. He was my delight; often it felt like he was my baby."

Roy agreed, noting, "Being the youngest, I guess I got babied a bit. Ruth has always said she was the one that raised me. Because Lewis was the oldest and I was the youngest, separated by 10 years and three sisters in between, I have always liked to call my brother and myself 'bookends.'"

One day the Card children had walked about a mile "down in the country" to play with their cousin, Donald Cobble. They went into the woods and were taking turns crossing a plank suspended across the limbs of two trees that stood closely together.

Selma, three years older than Roy, said, "We would walk out on the board and see who could knock the other off. Roy got pushed off and sprained his ankle. I had to carry him back to the house."

It was while they were engaged in a game of "Play like..." that Roy encountered another anxious moment. "He was about five or six years old, and we were playing like we were driving trucks. After a while Roy said it was time for us truck drivers to go to sleep, so we both climbed up a small tree in our yard. He said we should pretend to go to sleep – only Roy really did fall asleep," Selma recalled.

"He fell out of the tree and knocked the wind out of himself. It really scared us for a few minutes, but a man in our neighborhood happened to be home and rushed over. He was able to get Roy up and help him get his breath back."

Perhaps Roy's most dramatic moment came, of all places, while taking part in the weekend pastime – going to the movies. "We were at the theater and someone tripped him, causing him to fall through the glass at the entrance. His hand was cut pretty badly, but fortunately there was a physician's office next door and the doctor was able to treat him quickly before he lost much blood," remembered Ruth.

Despite the repeated flirtations with impending calamity, Roy was able to emerge from his childhood none the worse for wear. Even though he was only 13 when his family moved to Chattanooga, Roy had spent several summers working in Fort Payne mills, managing to avoid the typical industrial hazards those settings could present. Once he left Fort Payne, his affinity for near-disaster seemed to fade for the most part.

However, lessons about the virtues of hard work, gained through both practice and observation, remained deeply engrained in both Lewis and Roy, and continued to influence them throughout their lives. As the Card family's time in Fort Payne began to wind to a close, dedication to a strong work ethic would accompany them to Chattanooga. That philosophy would serve the Card brothers particularly well as they faced the uncertainties and challenges of blazing a manufacturing trail where no one had ever ventured before.

A look at the multi-generational Card family gathered for their reunion in 1931.

The Card sisters – Ruth Pickett, Selma McElhaney and Bett Bracken – are shown during an afternoon together in 2006.

3

Building a Cobble-Card Alliance

Long before brothers Horace and Henry Cobble considered moving to Fort Payne to take on managerial responsibilities in the hosiery industry, the Joseph H. Cobble family had established ties in a very different line of work. Joseph had worked both as a farmer and a sharecropper, spending many years toiling in the fields not far from the Chattanooga area, settling in the southeastern Tennessee community of Sweetwater.

There the Cobble brood was raised, including sons Horace, Henry, Albert, Joe and Roy, and daughters Cora, Anna Mae, Minnie and Ella. Anna Mae would become a primary link to the Card family through marriage. As it turned out, the Cobble brothers opted out of the demanding farming life and its limited financial rewards. While still living at home, they all were introduced to the hosiery trade. Albert used to tell about working in a hosiery mill at the age of 13, receiving the princely wage of 45 cents a day.

When they became old enough to launch out on their own, the lure of grander opportunities at the hosiery mills prompted several of the young, mechanically minded Cobble men to move to Fort Payne, with Henry and Horace eventually becoming mill operators.

Unlike their brothers, Albert and Joe Cobble never elected to move to Fort Payne. When they left the family homestead in Sweetwater, they chose a closer proximity, moving west approximately 50 miles to bustling, industrialized Chattanooga to pursue their livelihoods. But like their brothers, the initial career steps for Albert and Joe were to become "fixers" in hosiery mills, providing rapid response when machines broke down, replacing worn parts and making necessary adjustments to keep the equipment running smoothly.

In 1937, Albert and Joe diverted onto the entrepreneurial route and started Cobble Brothers Machinery at 315 W. Main Street, their original business site that consisted of only 1,500 square feet. At this machine shop, they repaired and rebuilt hosiery mill knitting equipment and made attachments for the machines.

They started this venture on the proverbial shoestring, borrowing $800 and working the first six months without generating enough revenue to pay themselves. Many times, "lunch" for the Cobble brothers consisted of a shared can of tuna fish. To attract customers, they even offered a delivery service, returning machines after they had worked on them.

The Cobble brothers' entrance into the tufting business actually came about by accident. In 1937, J.C. Wilson, who managed Colonial Coverlets in Chattanooga, brought them an order to make some parts for converting tabletop Singer 3115 industrial sewing machines into bedspread machines. Another frequent customer was George Muse, who asked the Cobbles to manufacture loopers, cutting parts and clutches for the revamped Singer machines used at his family's Muse Spread Company in Dalton.

The goose-necked Singer machines, along with the Union Special, were heavier than household sewing machines and designed for tailoring. These typically were used for producing bulkier textiles such as overalls and tents.

This single-needle Singer industrial sewing machine, displayed at the Shaw Research & Development Center, was adapted for tufting bedspreads.

In adapting the machines for making tufted bedspreads, a rocking or oscillating loop hook was designed to "pick" the yarn from a heavy duty needle, holding up to five loops at a time, while a corresponding cutting blade was added to cut the loops with a scissors action to create the individual chenille tufts.

Seeing considerable demand for this kind of work, the Cobbles began to expand their efforts in that area and in time decided to start making their own machines. Joe developed a loop cutting mechanism that worked very effectively, and they purchased modified Singer sewing heads to produce their own tabletop tufting machines.

Singer industrial sewing machines, refitted with multiple
needles for the bedspread tufting process in the 1930s, displayed
at the Shaw Research & Development Center.

With the supply of Singer 3115 machines available to be converted for tufting purposes starting to dwindle, Cobble Brothers – along with other machine shops – began to concentrate on building their own machines. This led to experimentation and the first tentative steps into tufting technology.

Brothers Decide to Part Ways

As brothers, Albert and Joe Cobble always maintained a strong relationship. In terms of both personality and general working styles, however, they were very different. Albert was married with a son, while Joe remained a bachelor for most of his life and was far more demanding of his employees. One machinist recalled that if he ever made a part that he knew was not perfect, he would take it to Albert, who typically would respond, "that'll do if it will work." He knew if he showed the part to Joe, however, he would make him dispose of it and try again until it was right.

At the same time, Joe was the more relationally oriented of the two. He was known for his generosity and eagerness to reward the diligence of his employees. Albert was more aloof, possibly because he had family obligations that competed for his attention, unlike his brother.

Jerry Hendricks, who has been involved in the tufting industry for more than five decades, commented, "I never really got close to Albert. He would come over and discuss things very quietly. He was never boisterous when you saw him. If he didn't agree with something you said, he would just respond, 'Hmmph,' and walk off."

As business partners, Albert and Joe often found their interactions strained because of their differences in personality and style. In early May 1940, the conflicts the brothers experienced at work reached a climax. They apparently had a major disagreement that could not be resolved, so both Albert and Joe approached longtime customer George Muse, offering him the opportunity to purchase one or the other's interest in Cobble Brothers. Muse elected to buy out Albert.

Although there are several versions of what instigated the actual dispute, one popular story is that the company needed 10 cast arms for machines and Joe ordered 100, perhaps caught up in a wave of optimism. On the face of it, this seemed like no reason to terminate a partnership, but it apparently was too much for Albert. He determined it was time to part with his brother professionally if they were to maintain peace as siblings – which they did.

After he and Joe decided to go their separate ways, Albert started Tennessee Textile Machine Company, where he made tabletop machines similar to those that were being made at Cobble Brothers. He eventually sold that company, next starting Super Tufter Machine Company, which would become a major competitor to Cobble in the carpet tufting machine industry.

One former employee observed, "There was a rivalry between the two brothers, but no animosity. They loved each other. They just couldn't work together."

Albert remained deeply involved in the development of tufting machines well into the 1950s and was credited for introducing some significant advances into the industry, including being among the first to produce a complete tufting machine model in 1943 or 1944. It was a "yardage machine," 45 inches wide with more than 100 needles and corresponding loopers, designed to make chenille housecoats, although cotton throw rugs were also produced on it.

In the 1950s he also introduced a dial attachment that allowed operators to adjust needle strokes through a radiused arm to select the desired stroke. This allowed for easier changes of the pile height and other tuning of the machine.

The Tufters Just Keep Going

Without his brother, Joe Cobble continued to guide his company in the development of tufting machines, aided by the support of Muse, who was from Sugar Valley, Georgia. Muse had been around tufting since the age of eight, when he worked for his father folding bedspreads after school. He had not intended to remain in the business, aspiring to a career in professional baseball. After seriously injuring his throwing arm in a car accident, however, his career with the Atlanta Crackers came to an abrupt end and he returned to the fast-evolving world of tufting and tufted products.

The Cobble Brothers Machinery Co. building that was located at 315 W. Main Street in Chattanooga. (Photo courtesy of Chattanooga-Hamilton County Bicentennial Library)

After the falling out of the Cobble brothers, Muse reportedly purchased Albert's share of the business for $5,600, producing $3,500 from his own savings and raising the remainder with the aid of his father. Muse's involvement with Cobble Brothers was primarily indirect, as a financial backer, and over time his investment in the company paid healthy dividends.

It has been said that a company's greatest resource is not its financial base nor its facilities, but its people – the talent, expertise, skills and enthusiasm they contribute toward achievement of the corporate mission and goals. In sports, there is a popular adage that "you win with people." In business, the "score" is kept on balance sheets rather than scoreboards, but the principle is the same. The alliance with Muse was significant for Cobble Brothers, but it was hardly the most critical personnel decision that company principal Joe would make.

In 1939, not long before he and Albert were to sever their business ties, Joe made a hiring choice that might have seemed like a simple gesture of kindness at the time, perhaps even a generous act of nepotism. In retrospect, however, it proved to be a pivotal milestone – not only for the future growth and success of the company (and subsequent companies) but also in shaping the course of the tufting industry.

Ready and Eager to Work

In recent decades, high school graduation has turned into a rite of passage for young people, marking completion of their compulsory education and commencement of the next chapter of their lives. For many students it has amounted to a requisite step for proceeding to college; for others it has meant the time has arrived to receive formal introduction to the real world of full-time employment. But either way, a hiatus of some length has often become the norm, particularly since the late-1900s.

"After all," many of today's young people would argue, "I've been working hard at school for 12 or 13 years, except for summer vacations. I deserve a break, a chance to rest before resuming my education or entering the work world."

For some of them, this might simply mean a trip to a favorite vacation spot. If they have enough money, it could involve a cruise with some friends. For those with the financial wherewithal, it could even entail getting on a jet, flying to somewhere like Europe and traveling around for a few months, possibly even a year.

For Joseph L. "Lewis" Card, however, in 1939 such notions were unimaginable, the farthest thing from his mind. In fact, the evening he graduated from DeKalb High School, near the foot of Lookout Mountain, he contemplated heading directly to Chattanooga immediately afterward to start his first job. His parents prevailed on him, however, convincing him to spend one last night in the family residence and then leave the following day to pursue his livelihood.

Of course it was a time deeply entrenched in the Great Depression. Young Lewis understood clearly that jobs during this era were a privilege, even a luxury, not just one of many options. "I was so eager, I wanted to hop on the train that night, but my parents insisted that I wait until the next morning. Because of the times we lived in, when we got a job we appreciated it. We knew we had better work hard to keep it. Not everybody could get a job, especially during the Depression.

"An 18-year-old back then was more responsible. The times were tough, but we didn't have to deal with drugs. We couldn't have afforded them anyway – nobody had any money in those days. It was a struggle for our mother just to give us a dime to go to the movies. It was a different world.

"Looking back on it, this probably gave us a lot of determination. We understood what could happen if we did not take care of ourselves and pursue opportunities. It was probably good for our generation – that's why (TV news anchor and author) Tom Brokaw called us 'The Greatest Generation.' The times became particularly hard starting in 1930, when I was nine years old (and the Depression set in)."

Lewis did not head north with vague ambitions, however. He had a plan. His uncle, Joe Cobble, had promised him a job upon graduation. True to his word, Joe had one waiting for him about 50 miles northeast of Fort Payne. It was time for heading to the "big city" and launching his career – whatever it would turn out to be.

Lewis did not have to concern himself about where to live. "Uncle Joe," a confirmed bachelor, was living at the Industrial YMCA on Mitchell Avenue in South Chattanooga (so named because it was nestled in the midst of the city's primary industrial district). "I roomed with him for a year and a half. Joe eventually got to where he was in good shape financially, and people kept asking him why he was still living at the YMCA. Finally he decided to go to the Read House downtown and rent an apartment there – he stayed for about three months, and then decided he liked living at the Industrial YMCA better and went back there. It was home to him."

A Mentor and a Role Model

Joe, 20 years older than his nephew, provided Lewis with far more than a job and a simple place to live. "He was my mentor – he gave me whatever advice I needed, as well as protection from the outside world," Lewis states. "He gave me my opportunity – and later on, my brother Roy's. He's the one that got us into this business and out of the hosiery business. He's truly the 'patriarch of the tufting machine.' There's no question that we have owed all of the success that we would one day experience in the industry to Joe Cobble, because he got us started."

As a mentor, Joe provided not only the wisdom and experience of an older man but also modeled strong, consistent character and a steadfast work ethic.

"His demeanor in working with me was the best thing ever," Lewis recalls. "He was extremely calm and trusting; he gave me confidence and independence, something you ought to teach all of your children. I suppose in a way I was like a son to him. He certainly was a father figure to me, particularly since we were involved in the same work. I'll never forget that when I wrecked his car, while I was still just a young kid, he said, 'Well (heck), it's just a car.'

"I never heard Joe express an unpleasant word. I never knew him as anything but an excellent relative. He had a good sense of humor, but could be serious when he wanted to be – or needed to be. And he was a very generous man. He was one of the finest fellows I ever knew. He stayed at his work a lot; being a bachelor, work was his life."

As one might expect, Lewis's introduction to work in the Cobble Brothers machine shop was anything but glamorous. He literally worked from the ground floor upward.

"My first job in the spring of 1939 was sweeping the floor, and then I started running lathes, milling machines and doing mechanical work. I picked things up from the other boys who worked there. There was no formal training. If you had a specific job to do, someone would show you how to do it. Joe and Albert would help us from time to time, too."

If someone had predicted that he would become one of the foremost innovators in tufting machine development, Lewis might have laughed at them on the spot. "I wouldn't say I was mechanical, although my father, Russell, was a fixer in the hosiery mill.

"I was not too involved in innovation before the end of the war. I don't know that I realized that I had any particular ability – even now. I guess I just learned by osmosis."

Incremental Progress, Needles at a Time

This "osmosis" included watching as Joe Cobble and his mechanics at Cobble Brothers began to advance tufting technology, one small step at a time. Tufting carpet was not even the slightest factor in anyone's imagination at that point. Machine tufting capabilities were crude, perhaps even primitive, compared to what they have become.

These men staffed the Cobble Brothers Machinery Co. in 1942.
Lewis Card is at the far left, and Joe Cobble stands at far right.

"When I started in the industry, there was a tufting machine – but it only had one needle in it. The single-needle machine had only been in existence for 5-6 years. B.J. Bandy and others were using it, but there had been no progress in the tufting industry," Lewis stated. "We took parts off the Singer 3115 and added new parts so we could do tufting with it. There were a lot of these used machines available in those days.

"Even though they were only making tufting machines with one needle when I arrived, everything from then on was a development of the same single-needle tufting process: multi-needle machines, pass machines, yardage machines," Lewis recalled.

The single-needle tufting machines were an improvement over hand-tufting, a process by which an individual sewed manually with a needle and cut with a pair of scissors. However, the machines still required painstaking labor – creating a line of tufts with the single mechanized needle, shifting the backing fabric over and then fashioning another tufted row.

"From the one-needle machines we started making a multiple needle machine to sew a tufted fabric by making a number of passes – we called it the pass machine. Each time we added a needle to the machine, we saved time. We could cut the amount of work required in half," Lewis commented. "We had a customer, Ben Winkler, who was in the bedspread business in the 1940s. One day he said to me, 'I'm tired of you coming down here and selling me a two-needle machine, then you come back three months later telling me you have a four-needle machine.'"

Adapting the Singer machines for multiple needles amounted to significant progress, but the goose-necked design of the equipment put a limit on what could be done with the expanded needle bar. "Then we started casting our own frames – they were all tabletop machines, and we adapted single-needle machines into multiple-needle machines."

Over time, Cobble Brothers was able to steadily expand the number of needles on the machine by multiples of two – to as many as 32 in the mid-1940s, covering four to six inches. Once the bar got that wide, however, they had to cast braces to support the added weight. To increase the number of needles beyond that point would require a totally different kind of construction, moving away from the tabletop-style, industrial sewing machine configuration.

Advent of the Yardage Machine

A major breakthrough was achieved as the United States' entrance into World War II loomed. In the early '40s, Cobble Brothers built the first so-called "yardage machine" for bedspreads, sewing up to nine feet in width. This was a freestanding machine, made up of two end frames and a set of lateral beams connecting the ends, capable of tufting bedspreads 108 inches wide, using 99 needles on 1 1/8-inch gauge (the spacing between each needle).

In his book, *Carpet Capital: The Rise of a New South Industry*, Randall L. Patton describes the precedent-setting invention: "Cobble's design had three purposes...he determined that if those ends were spread sufficiently apart and connected rigidly, the entire width of a bedspread backing might be passed between the ends and simultaneously stitched with parallel rows of tufts...if the cloth backing was passed over an ingenious roller that was embossed with helical ribs diverting from its center, a rotation of that roller would spread and hold the cloth taut over its entire width, maintaining the parallel configuration of the stitches...it would be possible to draw the cloth through the machine automatically by means of discs on which angular teeth were fashioned to engage it."

As momentous as it was, the arrival of this milestone was not shouted from the mountaintops surrounding Chattanooga. In fact, it was wrapped in a shroud of secrecy, so hush-hush that even veterans in the infant industry doubted rumors of its existence.

Rather than building the machine at the Cobble Brothers location, Joe Cobble instead chose to use the vacated old Lookout Mountain Hotel in Chattanooga. He had all of the first floor windows blackened to discourage industrial spies and curious passersby and built the machine in the lobby, using angle iron and channels. Since the machine was lubricated by grease on open connections rather than with oil, as would become the norm for future generations of machines, this machine – the first able to tuft an entire bedspread without making multiple passes – could not run very fast.

Five of these machines were built before the onset of America's involvement in the war, but none was sold commercially. One of them was transported to the old Velvetone Building in Calhoun, Georgia, owned by the Muse family, where it was set up and operated. Secrecy was again required since the company was not yet ready to market the machines. Windows of the building that housed it were boarded up and a guard stood by the only entrance.

Joe Cobble, Sid Manning, the shift manager and an excellent mechanic, and George Muse shared credit for the innovative machine, which became the forerunner of today's giant, high-tech carpet tufting machines. An application for a patent on the machine was filed in January of 1941 and granted on November 30, 1943.

Although Lewis Card was not as involved in this project as he would be in later years, he recalled the mix of exhilaration, determination and frustration that was involved. "We developed this nine-foot machine for bedspreads, but initially we had a lot of problems with it and had to go through a lot of changes and improvements to make it right. Nothing was ever easy. There just was no precedent, nothing to fall back on – we were out front, leading the pack."

Interrupted By the War

Experimentation on the tufting machine came to a virtual halt during the war as efforts were redirected to making parts for 75- and 90-millimeter artillery guns as subcontractors for the Wheland Company. Albert at Tenn-Tex and Joe at Cobble Brothers continued to dabble in tufting work a bit. However, because of the rationing of steel, they were extremely limited in what they could do with machine parts. The transition to participate in the war effort was not mandatory, but machine shops realistically made the shift to stay in business until peace could be restored.

Pausing during a dinner together in the 1940s are the Cobble brothers (from left) Henry, Albert, Joe, Roy and Horace.

Following the war, work on developing the machines resumed in earnest. "Most of our development in tufting started after World War II," Lewis stated.

It was a matter of taking the basic tufting concepts and expanding their capabilities, increasing the number of needles being used, reducing the gauge (spacing) between the needles, and adding new shifting mechanisms, controls and various attachments. Many of the veterans of the time described the experimentation during this period as "good ole trial and error."

It was a mind-stretching time for everyone, apparently even from a legal point of view. Lewis recalled Cobble Brothers applying for an initial patent and being informed by the patent office that what they had proposed could not possibly work. The only problem, he noted, was that it was already being done!

Passing the Baton

By the start of the war, Lewis had assumed a key leadership role at Cobble Brothers. Joe Cobble had relinquished his day-to-day control of the company in 1942, appointing Lewis manager of the shop. Joe retained his half-ownership of the business with Muse, but stepped away to explore other machinery-related interests and allowed Lewis to pursue his own ideas and innovations in the tufting machinery field.

With his first love being the hosiery business, having Lewis manage Cobble Brothers freed Joe to concentrate on the hosiery side of machine parts. There was no known reason for this preference, other than the hosiery business was what he was most familiar with, and it had retained his greatest interest. Nonetheless, Joe remained a trusted advisor and confidant for Lewis.

Max Marion Beasley, who joined Cobble Brothers in the late 1940s and continued to work closely with Lewis for more than 15 years, offered this observation: "Although I give credit to Lewis Card for the innovations that made Cobble great, I have to give credit to Joe Cobble for starting the company on the path which put it into this particular business and determined the product that made the Cobble name what it became in the tufting industry."

While Lewis was rapidly growing into his job and discovering new proficiencies as an inventor and innovator, not everything was a gain. Already the owner of a prosthetic eye, a shop incident also claimed the tip of one of his fingers. This occurred in 1942 while he was operating a punch press.

"I was flattening some parts and got to moving too fast. The parts were about an eighth of an inch thick, and the end of my finger was pulled off like peeling a grape. If the same thing were to happen today, I would probably still have the portion of my finger that was lost, but back then they didn't have the knowledge and technology to sew it back on."

This minor disability did not deter this ambitious young man, barely 22 years old. With wartime obligations behind them, carpet manufacturers and machine shops engaged in a frenzied battle to refine tufting machines for even greater uses than ever before.

One example was a race of sorts between Cobble Brothers and Cabin Crafts, a Dalton carpet maker. With the yardage machine suddenly becoming the tufting standard, both companies were striving to build a machine capable of inserting heavy carpet-grade yarn into a wide piece of backing material. By 1950 both companies had succeeded in their quest. Cabin Crafts built some of its own machines, but ceased production as Cobble Brothers rose to become a major supplier to the new carpet tufting industry in the Dalton district.

Instead, like many of its competitors, the carpet manufacturer utilized its mechanics to make their own unique refinements to machines they purchased to fit specific needs.

As the 1940s wound to a close, Lewis – like his longtime mentor – was on the verge of taking another momentous step for his company and the industry. Again this would involve a personnel decision that at first glance seemed inconsequential enough. And again it would be one that had a far-reaching impact, propelling tufting machines to greater heights than anyone could have ever dreamed. You could almost say this decision was "in the Cards."

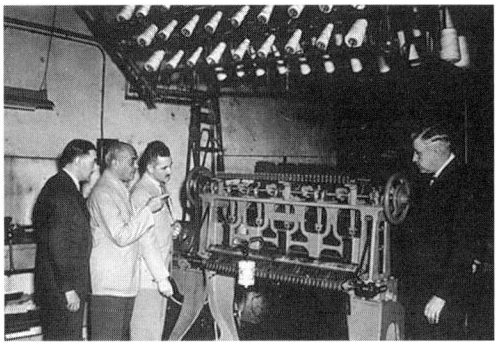

Potential buyers inspect an open-head tufting machine in the late 1940s. Albert Cobble stands at right.

4   A BOND BETWEEN BROTHERS

While Lewis Card and the tufting machine industry were establishing their hand-in-glove relationship in Chattanooga, his parents and siblings remained back in Fort Payne, continuing a small town lifestyle that revolved around the hosiery mills.

Lewis stayed in contact with the family, but his endeavors at the Cobble Brothers shop consumed his time six, sometimes seven days a week. So there were only infrequent occasions for casual family visits in Fort Payne. Several years later, however, when opportunity knocked again for Russell Card, the Card family followed older brother's path in returning to the Chattanooga area.

"My family moved back to Chattanooga when I was 13," Roy recalled. "My dad got a job at the TNT plant on Highway 58 in 1943 – it was wartime production in the latter stages of World War II. Dad always joked TNT stood for 'Too Near Tyner' (a suburban community at the edge of the city's east side). The manufacture of TNT explosives continued through the Vietnam War."

After his family relocated, Roy went through the Chattanooga public school system. He remembered what life was like, post-Fort Payne: "At first we lived in East Chattanooga, then we moved into an area called Glenwood – the property is now owned by Memorial Hospital. The house we used to live in isn't there anymore; now it's a hospital parking lot.

"Just below where the hospital is now on Glenwood Drive, right up the street from us, was the tuberculosis hospital where people went who had lung problems. I would walk to Hardy Junior High School, and later to Central High School, which was on Dodds Avenue then. Now that property is owned by McCallie School."

Like most boys of his era, Roy Card got an early start building his job resume. Many of his summer hours as an adolescent were spent in the Cobble-Muse hosiery mill in Chattanooga, working primarily in the dye house as a laborer, carrying specific quantities of socks in lots – unfinished socks on a big board – to the boarding room where the socks were put onto forms.

"The women working there would fuss with me if the socks were not suitable. From that point on, I determined that when I got a full-time job, I would work somewhere that I would not have ladies fussing at me," he quipped, reflecting on that time.

"The move to Chattanooga was pretty traumatic for me at the time – I told my mother that as soon as I graduated from high school I would go back to Fort Payne. But I never did. My uncle Henry Cobble, of course, owned and operated a hosiery mill in Fort Payne, and most of his children – and their children – still live there. My sister, Selma, eventually did move back to Fort Payne, and Ruth has lived in nearby Valley Head for many years.

"Compared to Fort Payne, going to Chattanooga was like moving into a big city. It was a major change for a 13-year-old. I knew no one in school or our neighborhood, and it took me several months to get acquainted with people and make friends. As a boy, I don't think I was real shy, but I wasn't real outgoing either – just kind of average."

Adventures with Gunpowder

Eventually, however, things changed for the better, Roy recollected. "By my high school years I had succeeded in making a lot of friends in the Glenwood area and we all played on the playground nearby. One year around the 4th of July, I suggested to a friend of mine, Niles Meacham – he was about my age – that we go to Stovall Hardware Store. There you could buy black gunpowder and we could make some rockets by putting the gunpowder into drinking straws. We bought about a half-pound of gunpowder, got some straws, put the gunpowder in them and started shooting them up into the air.

"After we had shot up several of our makeshift rockets, Niles fired one that sputtered away and quit. He went over to pick up the straw, but the gunpowder was still lit and blew up in his face. It blinded him temporarily, and I had to lead him to his house. He was in the hospital for about a week. I remember he used to have a lot of freckles on his face, but after the gunpowder incident, he didn't have many freckles anymore.

"The explosion didn't disfigure him or cause any permanent disability – that had been my greatest fear – but it really scared us. I felt very badly, especially since it was my idea that we go to get the gunpowder. Niles actually wound up playing football for Baylor School, and some for the University of Chattanooga."

This incident notwithstanding, Roy said he never regarded the experiment as evidence of any particular scientific or mechanical orientation. "I don't remember being mechanical as a boy – I fixed my bicycle when it needed it and that type of thing, and we all made homemade scooters and skates.

"We would take a piece of 2-by-4, nail a base and handle on it, take half of an old pair of roller skates and nail it onto each end of the board. And we made kites – we would find hollow, brush-like sticks in the woods and attach newspapers onto them, making paste out of flour and water. They were pretty lightweight and worked well. But most of us did things like that. I didn't do anything else mechanically that I can think of."

Neither was his academic performance, in his opinion, a harbinger that he one day would stand among the giants in the tufting industry. "I don't think I had any particular mechanical ability that manifested itself while I was in high school. I was not brilliant in school; I just passed."

As he was anticipating graduation from high school, unlike brother Lewis, Roy was not looking ahead toward settling into a vocation. He planned to enter the military, even though the possibility of a new war in Korea loomed ominously.

Bert Pickett, who married Ruth when Roy was nine, recalled the counsel he gave to his young brother-in-law. "When I got out of the service, the Korean War was just starting and Roy was talking about joining the Marines. When I heard that, I was jumping up and down, telling him not to do it. Ultimately, he joined the Coast Guard and not the Marines, and I had always thought I influenced him in not joining the Marines. But later I learned that Roy had gone to visit a friend in the Marines at Parris Island, South Carolina and afterward, he decided the Marines were not for him."

In point of fact, the decision was a bit more complicated than that.

As Roy recounted, "In high school I had joined the Marine Reserves. I guess I was naïve – or stupid – but when I joined the Coast Guard, I didn't get a discharge from the Marines. On my first leave, I went to the Marine Reserve center in Chattanooga and told the sergeant I was in the Coast Guard. 'You can't join the Coast Guard,' he told me. 'You're in the Marines!' But what was done was done."

The decision was a fortunate one, Roy noted. "As it turned out, during the Korean War the Marine unit I had been in was called up. A lot of those boys were wounded or killed. I was spared from having to be involved in any battles. Basically, in the Coast Guard the biggest challenge I faced was really being away from home for the first time. I guess it was good for me – it gives you an education."

Introduction to the World of Machines

Roy said he acquired his first useful, formalized mechanical training while in the Coast Guard, working with lathes in the engine room of the ship he was assigned to, learning to turn and sharpen metal, and also using tools that would cut the metal. Most of the machines were run by steam -- the manual labor involved monitoring and cleaning the mechanisms.

"I was really young when I went into the Coast Guard – I spent my 18th birthday in the North Atlantic on a ship. The reason I enlisted was that in those days they still had the draft, primarily by a lottery system. There was always a possibility of being drafted (into the Army). I had two friends with the last name of Cromwell who had joined the Coast Guard, and one had gotten stationed nearby at Lake Chickamauga, so I thought that would be pretty good duty for someone from Chattanooga. But that is not the way it worked out.

"The first year, my Coast Guard duties were on a ship that performed a weather patrol in the North Atlantic. It was called the Winnebago, named after one of the U.S. lakes, a commissioned ship that had been in mothballs since World War II in the Curtis Shipyard. I was involved in cleanup and welding, getting it seaworthy again. Our shakedown cruise was to the North Atlantic, and that happened to be over my 18th birthday. On a typical tour, we would stay out for 30 days, running in a 10-mile circle, putting up a weather balloon and coordinating with weather planes, and then we would come back. We would go past the Arctic Circle to chart icebergs in Baffin Bay, keeping a record of them unless they were small enough not to cause damage to any ship passing through the area.

"We saw a lot of ice and glaciers. Periodically we would hear a sound like a bomb going off – it was ice breaking off a glacier. Some of the icebergs looked four or five times larger than any manufacturing plant I had ever seen, and even then, we knew about 90 percent of each iceberg was underwater, out of sight.

"We were trained and equipped to provide air-sea rescue for the planes, if necessary, but that never occurred during my time of duty. Our cutter bounced around in the ocean like a matchstick.

"After that duty, we took a ship through the Panama Canal and went on to Hawaii, where I was stationed until being discharged. I was extremely fortunate, because I got out three weeks before the Korean War officially started. If I had been in the Coast Guard for three weeks longer, I probably would have had to remain in for the duration of the war. Overall I was glad to get that portion of my life behind me, and it did classify me as a veteran.

"Other than being away from home, there was no real downside. I was young and homesick, but didn't have to endure any major disappointments or dangerous situations."

Before long Roy would have to decide what to do with his life outside of a military environment. And again, good fortune, fate – or something higher – would intervene.

New Waves of Enthusiasm, Prosperity and Innovation

The 1950s provided an appropriate context for developments in the tufting machine industry, with numerous factors combining to change the fabric of virtually every aspect of American life and culture. In those years following the conclusion of World War II, the focus was shifted away from the war effort. A growing spirit of enthusiasm and optimism ushered in an unprecedented era of prosperity and new consumerism. This was bolstered by both major scientific discoveries and manufacturing advances.

On the political front, the decade had begun with Harry Truman in the White House, signing the peace treaty with Japan in 1951 to officially end the war, although war hero and retired Gen. Dwight Eisenhower served in the Presidency for most of the remaining years. Princess Elizabeth began her reign as Queen of England in 1952, at the age of 25. And Fidel Castro became dictator of Cuba in 1959.

A post-war building boom spurred huge growth in private home ownership, and by the close of the '50s, nearly 60 percent of Americans had acquired homes of their own. Televisions were popping up in households across the United States, with 90 percent of households having at least one TV by 1960.

McDonald's was founded in 1955, becoming the catalyst that initiated a phenomenal fast-food restaurant revolution, changing meal time routines and traditions forever.

Cartoons and animation became major news, with Walt Disney's celebrated "Cinderella" showing in movie theaters for the first time in 1950, the same year that Charles Schulz's beloved "Peanuts" comic strip was being introduced. Other entertainment highlights included the grand opening of Disneyland in 1955; the first performance by a shy country boy from Tupelo, Mississippi named Elvis Presley on the "Ed Sullivan Show" in 1956; and the signature debuts of "The Sound of Music" (on Broadway) and Barbie dolls, both in 1959.

On the science front, the Salk polio vaccine was introduced in 1952, and DNA was discovered by scientists James Watson and Francis Crick in 1953, the same year DuPont began commercial production of a synthetic fiber called Dacron, a material that at one point would be used in the manufacture of carpet. The Russians launched the so-called "space race" in 1957 by putting Sputnik into orbit, and the microchip was invented in 1959, although its full impact on the world of computers and automation would not be felt until decades later.

Appropriately enough for the geographic hub of tufting machine innovations, songs topping the musical charts during the decade of the '50s included Patti Page's "Tennessee Waltz" and "Chattanoogie Shoe Shine Boy" by Red Foley.

And a footnote to the 1950s, one that would take on significance for the world of tufting machines about 30 years later, involved the births of Charles Monroe (in 1950) and Lewis "Lewie" Card, Jr. (in 1952). As Lewie would one day observe, the second generation in this story of the carpet tufting machine industry was barely getting started just as the third generation was being born.

Plans Shuffle for the Cards

After Roy was discharged from military service in the summer of 1950, he returned to Chattanooga and began looking for a job. "I applied for a job at DuPont, one of the major employers in the city at the time. I went through the hiring process and they said they would hire me," he recalled. "But then I was sent to a doctor for a routine physical. I had just received an honorable discharge from the Coast Guard and anticipated no problem, but the doctor discovered I had a slight curvature of my spine, so DuPont couldn't hire me.

"It was disappointing. I was just looking to start working somewhere, anything that would pay minimum wage, which back then was about 75 cents an hour. It would have been an entry-level job, but at least I would have had my foot in the door at DuPont and could have tried to work up the ladder there."

In retrospect, Roy's failure to pass his physical represents one of the possible answers to the recurring question of why seemingly bad things happen to good people: Because sometimes a better, preferable alternative lies ahead, waiting in the future.

"If I had passed that physical at DuPont," he admitted, "I probably would have retired at DuPont. I would not have changed jobs (because of the value of working for an established, stable company). Back then, if you could find something to do, no matter what it was, you usually stayed there."

Once the door to DuPont slammed shut, however, Roy was forced to resort to plan B – even though he really did not have a backup strategy.

"I talked to a man with a service station where my Mom and Dad traded and he offered me a job. I don't remember what the job would have been.

But when I told Lewis, he said I didn't have to work there. 'I can beat that. Come to Cobble Brothers,' he said, 'and we'll find something for you to do.' Working in a service station was the bottom rung of the career ladder back then – there was no such thing as a self-service station – so I was glad to have an opportunity to work with my brother, whatever they had for me to do."

Lewis's offer did not come about because of a close-knit bond between the brothers. It was a simple case of family loyalty. "When I was seven or eight years old, he had left to go to work for Cobble Brothers. Growing up, because of the gap in our ages, he and I had not been close. There was too much time between us – he was a teenager before I got to know him much at all – and I'm sure he really didn't want a little brother tagging along after him then," Roy said.

In subsequent years, bringing Roy on at Cobble Brothers would prove to be the most significant hiring decision Lewis ever made. And with the passage of time, the age gap ceased to matter. A true bond was forged between them – not only a bond of brothers but also that of peers, two men with corresponding, complementary talents that would unite to revolutionize an industry.

As so often happens, taking a backward glance through time offers wonderful perspective. "Now we're really close. We worked together for some 50 years, and I think just being able to get along is pretty amazing. We grew closer as the years went by," Roy commented.

Many years later, Lewis conveyed his full awareness of how fortuitous his brother's failure to get on the DuPont payroll proved to be. "Yes, if he had gotten on at DuPont, I would not have had anybody to help me. I probably would not have made it without Roy's help."

In his typical low-key, self-effacing manner, however, Roy downplayed the countless contributions he was able to make through the years. "Well, it was nothing special that I did. I was just in the right place at the right time. I had no great thoughts about going into this business, doing this and that, and moving on from there. After all, when I arrived, the main business was not even tufting equipment. It was making parts for the war effort – we were still subcontracting for Wheland and Watervaleet, a government arsenal in upstate New York, making gun parts, firing pins for the big guns, the nine-millimeter guns."

Gaining Experience in the Machine Shop

Initially, Roy spent a brief time cleaning floors and performing other menial tasks. But soon he was toiling in the machine shop at Cobble Brothers, doing drill press work, running lathes and other machines, and making parts for yarn creels. These creels were designed to hold cones of yarn being fed into the tufting machine. The machines at the time were 60-120 inches wide, but still fairly primitive, being used primarily to produce simple bedspreads and rugs.

"I did that for two or three years, before I moved to the sample department in 1953, where I was responsible for running samples for carpet manufacturers. I continued working there when we moved our facilities to Riverside Drive in 1955," Roy stated.

"Early on, I was just glad to be on board, having a job that paid a decent wage. In the sample department I enjoyed being around tufting machines and learning what they did. I also liked working with the customers and being able to satisfy their needs.

"Most of the time customers were there when we were running samples. They would tell us they wanted to be able to run a heavier yarn for their carpet – weight was always a big item. Sometimes I would tell them it seemed as if they were selling carpet like meat – by the pound."

Roy observed that no one in the fledgling industry had any lofty illusions of what it would eventually become. It was simply a matter of working one day at a time, meeting challenges and overcoming obstacles. "Nobody had any dreams that the (tufting) industry would grow as it did – it was beyond anybody's imagination; certainly beyond mine. At the time, we were struggling just to keep bedspread and rug machines running as they should."

Just as Joe Cobble had served as a role model and mentor for Lewis, giving him exposure to the overall operation and then extending to him the latitude to utilize his gifts and abilities in a hands-on manner, Lewis slowly began to pass on what he had learned to his younger brother.

As Lamar Card, Lewis's oldest son, recently observed, "Joe Cobble saw that Dad had the ability and gave him the opportunity to put it to use. Dad combined that ingenuity with a tremendous work ethic. He was just a bright individual, teaching himself what a lot of people would spend years studying in college. I don't know where he got it, but I have always marveled at how Dad was able to achieve such a grasp of mechanical physics in the way that engineers did, only he was self-taught.

"He later taught Roy – or rather, showed him – the same thing, how to see things differently. Over time Roy developed the same sense and they became a true team of inventors, an ideal situation for working together in a very narrow area (of enterprise)."

Gradually the brothers came to appreciate the fact that they were blessed with an unusual synergy, combining their innate inventiveness with separate strengths that would enable Cobble Brothers – and later companies – to set the pace in tufting innovation and initiative.

Lewis's emphasis was on the mechanism, building the machines, while also being able to flourish on the business side, forging strong relationships with financial institutions and suppliers, making sure the company was able to remain profitable. Roy's specialties included the application of innovations they developed together, combining mechanisms to create new and distinctive tufting effects, and also giving ample care and attention to valued customers.

Speaking on how they were able to share leadership responsibilities, Lewis said, "We were able to do it (work together) without throwing our weight around because of our understanding of each other."

Roy added, "We were just in the business for survival. We weren't trying to prove anything to each other – or to anybody else for that matter. Why or how we got along so well, working together day after day for so many years, I don't know."

But Lewis quickly responded, "I do. I probably would not have gotten along with somebody who was not like Roy."

Another trait the men shared was an uncompromised devotion to their trade. Neither Lewis nor Roy had many interests outside of their tufting work. Roy summed up their philosophy: "You do more for your community by providing stable employment, and then you can let the employees go out and become personally involved in community affairs."

Time for Family – After Near-Tragedy

That is not to say that the Card brothers lacked any personal life away from the machine shop. Roy and the former Virginia Cade married in August 1951, although the event was delayed more than half a year following a near-tragic car accident.

They had planned to marry in December 1950, but two weeks before the wedding day, the mishap occurred while Roy was driving Virginia back to her mother's home in North Chattanooga. They were traveling on 3rd Street when another vehicle turned into the path of his car, striking and pushing it into a telephone pole. Cars did not have standard seatbelts back then, and Virginia's head struck the dashboard hard, crushing her jaw. Following surgery, her jaw had to be wired shut for six months, which made it difficult for her to receive proper nutrition, but eventually Virginia made a full recovery.

Fortunately, Roy's injuries were far less serious. He had been knocked unconscious by the impact, but suffered only a head bump and no broken bones, and was released from a hospital the next morning.

Roy and Virginia eventually would have three children and nine grandchildren. Daughter Renee married Charlie Monroe, who would become a notable part of the tufting machine saga. Sons Brian, married to Terry, and Brad, whose wife is Paula, also would grow up to become participants in the family business.

As for Lewis, in 1941 he married Katherine Maxwell and they had three children, who presented them with seven grandchildren: Lamar, a film producer in California, who is married to Sarah; Janice (married to Danny Henderson), who has served on the board of directors for the family business; and Lewis, Jr., whose wife Margaret died in 2008. He married Becky Hatfield in 2010. Katherine passed away in 1975, and Lewis, Sr. married again in 1976 to Rocelia Owens Benton.

Innovation through Problem-Solving

For the most part, both Lewis and Roy devoted their waking hours to exploring and expanding the potentialities of tufting machinery. One step at a time, they and the small team of men around them devised and introduced minor improvements to enhance the capabilities of their machines. These changes, however, were often motivated more by problem-solving than pure innovation.

Slowly they were able to expand the size of the rug machine a number of times, to six-foot, nine-foot, 12-foot and 15-foot widths. "Every time we increased the width of a machine, we had to increase the weight and strength of the machine," Lewis pointed out.

"Then we were using 5/64-gauge for loop pile, meaning there were 12.8 needles per inch in the tufting machine. That was a pretty fine gauge for that time. If we could keep the tolerances of our machines to 50-60 one-thousandths of an inch, we thought we were doing good."

The Cobble brothers, shown in the 1950s, are (from left) Joe, Henry, Horace, Roy and Albert.

As Roy began taking an increasingly significant role at Cobble Brothers, the direct presence of Joe Cobble correspondingly diminished. He remained available for consulting well into the 1950s, but confident that the tufting machine company was in good hands, Joe was content to concentrate on other projects, especially the hosiery business. Even though he was withdrawing his hands-on involvement at Cobble, his influence left a lasting impact.

"He was easy to get along with, very helpful to me – compassionate and generous, more than we ever knew. He took joy in helping people when they needed it – family, friends, and people who worked for Cobble," Lewis said.

Max Beasley, a longtime Cobble employee, agreed with that assessment in personal memoirs he compiled about that period. (Beasley and Lewis had known each other as boys in Fort Payne, and Lewis initially hired him to do some design work in 1946.) Commenting on Joe Cobble, Beasley wrote, "… Joe always was one to share his good fortune, and was a very generous man. Business had been good, and several years in a row he chartered a Greyhound bus and took the whole shop to Panama City to go deep sea fishing."

Beasley also noted that Joe was particularly lavish in his generosity at Christmas, giving employees substantial bonuses – sometimes far beyond the norm for that era – as a reward for their hard work and dedicated contributions to the company's success.

At the same time, Joe Cobble was not one to call attention to himself. In a crowded room, he might not have stood out at all to anyone that did not know him, even though at 6-foot-1 and being heavyset, he was a good-sized man for that time. "He was an average person, literally an average Joe. He didn't have any unique, distinctive traits that I recall," Lewis commented.

Because he was older than Roy, Lewis enjoyed a much closer relationship with Joe. In addition to their family ties, the two men had been roommates, and Joe had served as boss and mentor, confidant and counselor for his nephew.

While Roy held great respect and admiration for Joe, their relationship was not nearly as collegial or close-knit. "Joe was very direct with me. If he said it, that was it. Just a straight-forward kind of person, as far as I was concerned," recalled Roy.

Reflecting back on his beloved uncle, Lewis summed up Joe's greatest contribution to his and Roy's lives. "The best thing about Joe is that he gave us the opportunity to get out of the hosiery mill business."

Tough Projections for the Tufting World

Not everyone at that time would have agreed that getting out of the hosiery mills was such a good idea. Despite the unrelenting wave of optimism that dominated many industries during the 1950s, the tufting industry collected more than its share of naysayers.

Around the time Lewis and Roy Card were teaming together and starting to hit their stride, some veterans of the carpet industry were offering woeful pronouncements about the future for tufting. One old-line carpet executive in Philadelphia described tufted carpet in 1952 as "glorified bedspreads." Admittedly, for a while the quality of the material was not much more than that, although this would soon change. Another notable authority was quoted widely in 1956 as declaring that the tufting industry had "finally peaked out."

Participants at the Tufted Textile Manufacturers Association meeting in 1948.

Such pessimistic prognostications, however, were generally limited to a small minority that lacked both vision and imagination, and such negative appraisals soon vanished. By 1954 it became evident to most knowledgeable observers that tufting held the future for carpet. While carpet weaving looms were hard-pressed to produce a mere 100 yards of carpet in 24 hours (approximately four yards an hour), 300-500 yards of carpet could be tufted in a 24-hour time span (as many as 20 yards per hour). And as the post-war building boom began to place much higher demands on the floor covering business, the decision to shift carpet manufacturing's focus toward tufted products was motivated as much by expedience as it was by profit.

That does not mean the tufting practitioners had it all figured out. Not by a long shot. As Lewis has frequently observed about the industry climate of the early 1950s, "We knew little about the tufting machine – and our customers (the manufacturers of carpets and bedspreads) knew even less." About the only thing that all the interested parties did know for certain was that with each succeeding year, demand seemed to exceed supply in the carpet industry.

Innovations at Cobble Brothers continued, resulting in new and better ways for making tufted carpet. In response, the region's new carpet moguls took delight in investing in these increasingly faster and more efficient machines, looking with great anticipation toward the future.

That kept Lewis, Roy and their employees in a relentless pursuit of machinery that was bigger, better, ever faster, still more efficient and endowed with greater precision. And in time, their efforts would take them far beyond the friendly, familiar environs of Chattanooga and northern Georgia.

5 DREAMING BIGGER THAN THEY COULD SEE

"Timing is everything," a popular motto informs us. For Lewis and Roy Card and Cobble Brothers, timing indeed proved a perfect ally for their special combination of hard work, determination, mechanical expertise and innovation. One reason the detractors of carpet tufting in the early 1950s were so wrong in their dismal predictions was a failure to take into account the phenomenal confluence of factors that would propel the industry to unimagined success.

Prior to World War II, the tufting industry was very simple and straight-forward. As Thomas M. Deaton writes in *Bedspreads to Broadloom,* "All it took was a single needle machine, women to sew and a building." That suddenly changed, however, after the final curtains dropped on both the European and Asian theaters of World War II.

With signatures barely dry on peace treaties marking the cessation of wartime hostilities, Americans from coast to coast began to enjoy a sharp increase in per capita income. This, in turn, brought an unprecedented wave of consumerism throughout the United States. Home ownership soared to all-time highs. As advances in tufting technology made carpet economically accessible to many families for the first time, demand for soft floor coverings rose dramatically. Savvy marketing convinced novice homeowners of the advantages of carpet as an attractive alternative to hardwood flooring. Before long, reasonably priced "wall-to-wall carpet" had moved from being a privilege to a preference.

As consumer demand for many products escalated, so did reliance on mass production. This applied to old-line carpet manufacturers who gradually shifted their focus and energy from time-consuming weaving practices to the more expeditious tufting process. Evolving machine technology, merging with scientific discoveries that resulted in attractive, durable synthetic fibers, further revolutionized carpet for in-home and commercial purposes in terms of both quality and affordability.

Other significant advances included use of basic printing and dyeing techniques, and development of new and better materials to serve as backings for carpet. Two other factors, eventual expansion into new carpet markets – including industrial, outdoor and automotive applications – and the arrival of proven leadership and experience of veterans from the world of woven carpet migrating to the domain of tufting, helped to further solidify the industry.

Extremely Conducive Environment

These factors and influences combined to create a very favorable environment for men like the Cobbles and Cards to practice and perfect their craft. And for someone like Lewis Card, a general attitude of "can't lose" and "nothing ventured, nothing gained" also helped.

Lewis often reflected on a statement made by his longtime mentor, Joe Cobble, that empowered him to explore the limits of tufting technology.

"One time when we were having some real difficulties, financially and otherwise, and were considering doing something risky, Uncle Joe told me, 'Son, we came here with our (butts) and a hat, and we can always leave with that.' In other words, we arrived here with nothing, and when we got out, the worst that could happen was that we would be in the same position as when we started.

"Taking this attitude gives a lot of latitude. He made this statement when we were not too big and didn't have a whole lot to lose – he was trying to keep me from worrying too much about things. If you're not worrying about something you can't control anyway, it enables you to venture out and take risks.

"You have to take risks in developing new machines and innovations. There is no telling how much money we spent on things that we threw into the junk pile. You can't worry about trying things that don't work. We weren't engineers and much of the time we didn't know if something would work or not. But we were sure going to find out. I wouldn't say that is the best way to go about it, but it was the way we had to do it at the time."

Meanwhile, venturesome industrialists, accustomed to taking a big picture approach to business, were eager to respond to the growing fervor for carpet and capitalize on its possibilities.

For instance, Robert Shaw, founder of Shaw Industries, had his own explanation for the steady, incremental growth of the carpet industry: "I am a great believer that when you put the building blocks together and get up on the building blocks, you can see a little further. Then the horizon gets bigger every time you put another building block up."

Shaw conceded, however, that innovators and entrepreneurs should not feel limited even by those building blocks. They also had to cultivate a bit of the dreamer in themselves, a willingness to "think or dream bigger than you can see," he said.

This trait probably contributed to the achievements of the Card brothers and their companies. However, they were cautious never to let their ideas become too grandiose. The view from the machine shop, after all, was a bit different from that of the carpet mill.

'Only One or Two More Years...'

Max Beasley, whom Lewis initially hired as a part-time draftsman and illustrator, remained involved with tufting machinery for more than 20 years. He contributed significantly to a number of mechanical advances while in Lewis's employ. In later years, Beasley wrote memoirs of his life and experiences, compiling an amateur historian's view of the period and the industry. He offered this machine shop perspective:

"Nobody outside the business visualized what carpet production could do. What carpet was available (at first) was so inferior that nobody could see that carpet could ever be made that would be worth buying... Only these Cobbles and their employees had faith in the future. Many times I have heard someone say, 'One or two more years and that'll be it for the tufting industry.' It never happened!"

Lewis expressed it a little differently: "We took it day by day. We couldn't see into the future, and didn't know of anybody that could. We believed in the business, convinced that it had great potential for new development, so we just continued to look at it positively without worrying about what the future held."

In describing the progressive development of the tufting machine, Beasley liked using the term "incorporeal," which suggests a lack of material substance or a clearly defined future. The machine's conception and development, as he termed it, were daily being "discovered and wrestled into reality."

Reflecting back on those early, uncertain years, Lewis agreed. "Our key to survival was continual mechanical innovations and new development. So I guess in that sense, yes, the tufting machine was incorporeal."

Then he reiterated the greatest advantage that he, Cobble Brothers and other innovators enjoyed, as cited earlier: "At the outset, we knew little about the tufting machine, but the good news for us was that our customers knew even less."

Principles Remain Constant

Interestingly, although the tufting machine was a continual work in progress through the 1950s – and remains so even today – its operative principles never changed. As the Cards worked at expanding the size of their machines, increasing the number of needles and other gauge parts, as well as learning how to make the equipment faster and more efficient, they essentially remained "sewing machines," although far more sophisticated.

Throughout the steps of development and innovation, the process still consisted of inserting a specific face yarn into a pre-woven backing by needles, having it caught by reciprocating loopers, and then in a later process securing the inserted tufts into place by the blooming effect of the yarn or the addition of a secondary backing.

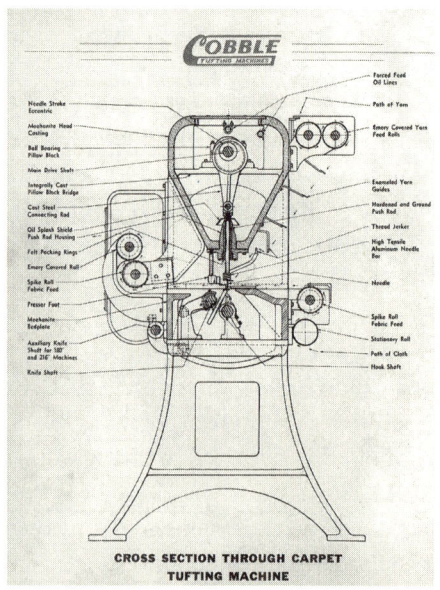

A cross-section view of a carpet tufting machine
produced by Cobble Brothers in 1956.

Easily explained – but not nearly so easily achieved. For that reason, every incremental improvement represented a small triumph, a significant stride toward the realization of machines that one day would make carpet a standard fixture in most homes, businesses and institutions, many recreational areas, and even motor vehicles.

In retrospect, this time of experimentation and discovery came to be known as "the Golden Age of Development." As Deaton writes in *Bedspreads to Broadloom,* "Virtually all technology which lifted products out of the category of washable cut rugs into a viable and cheaper alternative to woven carpet was put into place. The conversion to synthetic yarns, the development of basic printing techniques, the addition of (secondary) backing, and new machines evolved onto the scene. Floor covering engineers, chemists and machinists made all sorts of efforts to make manufacturing 'cleaner and quicker' so that carpet has become an affordable luxury product, but the basics for future design were in place by 1960."

Critical Shift from Grease to Oil

One of the keys to increasing the tufting machine's speed involved a change in how moving parts were lubricated. In the first yardage machines developed in 1940, portions remained open so movable parts could be greased as needed. By 1949, however, the first fully enclosed tufting machine had been developed, with oil as the lubricant for the working parts. The immediate effect was a substantial increase in the operating speed of the machine.

Initially the enclosed machines were used to produce bathmats, but not long afterward larger rugs and carpets followed. In 1952, a major step forward was taken with the introduction of a 12-foot wide machine with the capability of producing loop pile in 5/32-inch gauge (6.4 needles per inch) capabilities. The next year, cut pile was added in 3/16-inch gauge (5.33 needles per inch) and an 18-foot machine was introduced. Even though carpet was made in 15-foot widths, the additional machine width was necessary to accommodate backings – then consisting of cotton – that would shrink up to 20 percent in the finishing process.

The ability to produce room-sized carpet was a notable advance, but there remained a huge quality issue. Lewis vividly recalled the first carpet his machines produced, admitting they fell far short of striking competitive fear in the hearts of woven carpet manufacturers.

Those first samples were, "like a glorified bedspread. It was made on cotton duck (material similar to canvas)," he said. "The quality of carpet was an evolution, just as the development of the tufting machine was an evolution. The first products we were able to make in the new business, frankly, were not that good compared with where we would eventually get to – it took a lot of effort on the part of bedspread manufacturers to get the carpet business going. It was truly bedspread manufacturers who were the ones that started the carpet industry."

Moving from bedspreads to rugs to carpet was a slow progression, one that required patience and perseverance, according to Lewis. "The whole thing was about being in the right place at the right time. And being lucky enough to be the one that was there then!"

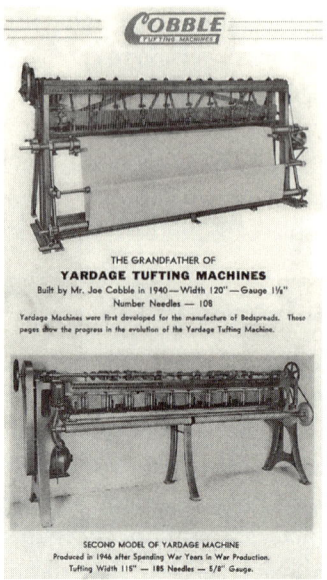

This bedspread yardage tufting machine was a precursor of the carpet yardage machine that would be developed about a decade later.

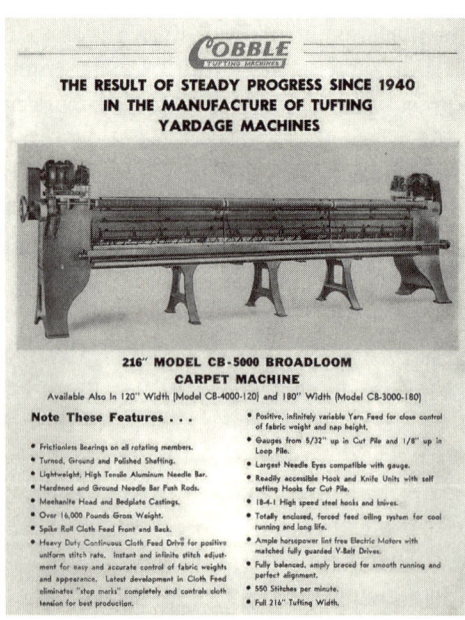

This Cobble Brothers Machinery Co. catalog page shows a 216-inch broadloom carpet machine produced in the mid-1950s.

Bedspreads, Bathmats and Robes...to Carpet

While considerable time was being devoted to learning how to replicate the quality and appearance of woven carpet through the tufting process, much of the effort in the early 1950s remained concentrated on producing other tufted goods. Demand for these products – bedspreads, bathmats and scatter rugs, and robes – had also increased exponentially as a result of the post-war consumer boom.

The Tufted Textile Manufacturers Association reported that tufted bedspreads went from five million produced in 1940 to more than 19 million in 1951. Production of tufted bathmats, scatter rugs and carpets combined to grow from 1.2 million in 1939 to nearly 41 million in 1951, and tufted robe production rose from 860,000 in 1939 to nearly 11 million in 1951.

So as Lewis and Roy Card, Cobble Brothers and other tufting machine innovators continued to explore the possibilities for tufted carpet, the manufacture of machinery to make other tufted products remained their bread and butter as the new decade dawned.

"In terms of our production, we were making both kinds of machines to start with, but early in the 1950s we were able to shift our primary emphasis to machines for the carpet industry," Lewis said.

Despite substantial post-war growth in demand for various types of tufted products, consumer preferences eventually changed. As a result, all tufting enterprises began to concentrate their energies on the development and refinement of carpet. In the early 1950s, production of carpet and large-sized rugs rapidly began to eclipse the production of bedspreads, bathmats and robes as the popularity of those textiles started to wane. The manufacture of small tufted goods never exceeded $100 million annually, according to Deaton in *Bedspreads to Broadloom,* but steady growth in tufted carpet would push that industry past the $1 billion mark in the early 1960s.

As one might expect, this major manufacturing shift had a sweeping impact, even on the communities that served as home for the industry. As John Longwith writes in recounting the history of the Dixie Group (formerly Dixie Yarns) in *Bound and Determined,* "Dalton's transition from bedspread junction to carpet crossroad began in the early 1950s, when the developments of nylon and tufting machinery gave birth to a new industry geared to the mass production of carpet. Before then, the idea of wall-to-wall carpeting would have seemed outlandish to most homeowners... But nylon tufting drove carpet prices down, making wall-to-wall carpeting a low-cost alternative to wood flooring and doing for the carpet industry what Ford's Model A had done for the automobile industry."

Short-Lived Secrets

When tufting machine manufacturers were developing and testing their innovations, every effort was made to maintain secrecy. Competition was intense and everyone was seeking an advantage in this highly specialized market. No matter how closely information was guarded, however, secrecy achieved only limited success. "Industrial piracy (of concepts and design) was not uncommon," Lewis pointed out. "When we shipped a machine out, we lost our experimental work and ideas. There are really no secrets in the tufting machine industry. When it goes out your door, you've lost it."

An early open-head yardage bedspread machine displayed
at the Shaw Research & Development Center.

One approach to protect new developments was to seek patents for them. During their careers, Lewis and Roy were awarded well over 100 patents each in the United States and internationally, but every patent involved a lengthy process, often taking up to three years or more. In the meantime, competitors could be duplicating innovations and implementing them on their machines. And the uniqueness of the tufting machine industry often confounded even the patent offices.

Lewis commented regarding the universal pattern attachment, one of the first breakthrough innovations: "When we first started using the universal pattern attachment, there were no experts in tufting then, even ourselves, let alone the U.S. Patent Office. The industry was so new that no one had been around long enough to become an expert – including ourselves. We were just out there working hard, trying to get something done."

And "getting something done" was exactly what Lewis, Roy and the assembled cast of tradesmen at Cobble Brothers Machine Company were doing. The universal pattern attachment, developed in the early 1950s, became a classic example.

This device controlled the yarn across the full width of the tufting machine, enabling it to produce carved effects that corresponded with a design that had been fixed on a pattern drum. In appearance and effect, the attachment operated according to principles similar to the roll of a player piano that dictated which keys would be played on it and when.

As Randall Patton explains in his book, *Carpet Capital,* "(Lewis Card) conceived a means by which to engage and disengage individual needles from the reciprocating needle bar so that a needle could insert a tuft only when required. This controlled needle technique anticipated some of the refined computer-controlled specialty machines now in use."

Finding Help In-house

Sharing some of his brother's attributes, Roy advanced rapidly at the small shop, taking on added responsibilities and acquiring new skills. He learned how to run equipment like the screw machine, then graduated to the experimental or "sample department" as technical services manager, running smaller tufting machines to develop new fabrics as well as interfacing with customers.

In this role he studied and mastered the complexities of machines and gauge parts interacting with yarn and fabric. Lewis was leading the innovative charge, with the aid of numerous key individuals; before long Roy became the most valuable of them all.

He and Lewis worked well together, often collaborating to resolve tufting dilemmas and investigate new possibilities. At the same time, their complementary strengths enabled them to serve Cobble Brothers and their succeeding companies in different ways, from both business and mechanical standpoints.

Despite the age difference that had precluded their growing close as brothers in Fort Payne, it was not long until the working relationship between Lewis and Roy developed into one of mutual admiration, love and respect.

"His work habits were always excellent, and he's even-tempered," Lewis stated, smiling as he added a qualifying quip, "but not like a lady I knew once whom people described as even-tempered only because she was mad all the time!

"Roy has always been easy to get along with, level-headed, and doesn't let himself get in the way of anything. He's probably the best man I ever knew – he and my Uncle Joe. Roy was a good man to go to and bounce ideas off, whether it was work, a problem you were trying to solve, your personal life, etc. That still happens today, in fact."

Roy's comments about his brother were equally praiseworthy. "Lewis was intense, and always dependable. You never asked him to do anything that he didn't follow through. If there was anything I wanted him to do, he'd do it for me.

"As far as being a businessman, he was brilliant – a true leader. He had the ability to lead people, not necessarily to do his bidding, but to do what was right – what needed to be done.

"I definitely would not have been in this business without him, and would not have done as well as I have without him. Neither of us thinks we're always right about things – we made a lot of mistakes along the way. I'd feel covered up about problems I had, and Lewis's response always was, 'Take your licking and just move on.'"

As gifted mechanically as Lewis was, whether rolling up his sleeves in the shop or pondering a recurring machine problem while sitting in his office, he was equally astute in the operational side of the company. He was comfortable negotiating with bankers, presiding over company growth strategies, and conceiving unique approaches to stimulate machine sales.

Roy was adept at discerning the best applications for new machines and attachments, while at the same time establishing a strong rapport with customers, learning what they would like their machines to be able to do, and then searching for ways to get those things accomplished.

Even though Lewis had seniority, both in terms of age and years in the industry, the issue of "who's the boss" never mattered to either of them, they said. And unlike the Cobble brothers, their differences never became deep or severe enough to cause a business rift.

"All we ever wanted to do was to get the job done, whatever it required. We had our serious discussions and disagreements, but never such that we split up. It's like arguing with your wife. Just because you disagree doesn't mean you're going to get a divorce," Lewis said.

The Ever-Present Uncle Joe

Meanwhile bachelor Joe Cobble, not marrying until late in life, continued to regard his simple room at the Chattanooga YMCA as "home." And long after he had stepped away from Cobble Brothers to devote his attention to the hosiery industry and other interests, he remained a ready resource for Lewis.

Since his evening hours were generally free, Joe often would lock up his offices at the knitting mill he had established with George Muse and drive to Cobble Brothers to see what was transpiring there. These visits were a combination of business and pleasure, as he readily responded to opportunities for offering counsel to Lewis.

The nephew regarded Joe as "a shoulder to lean on," always available to listen. Joe would advise his nephew, seeking to calm his fears and ease the day-to-day pressures of operating a business, but was determined not to foster a dependency. Like a mother bird, Joe understood when the time was right to force his "chick" to fly on his own.

"I remember many situations that he used to teach me to be independent. One time after he had put me in charge of Cobble Brothers, I came to him to talk about money for a new project. He said he would not put any money into it, but if I could find somebody – a bank – to loan the money, then 'go get it.'"

This philosophy proved extremely wise soon after the company instituted plans in 1954 to merge all of its separate operations under one roof on Riverside Drive. More than 80 Cobble Brothers workers had been spread among six satellite locations around Chattanooga, including sites on East and West Main streets, and it was determined that bringing everyone together would be a better use of time, resources and manpower.

A view of the Cobble Brothers Machinery Co. when it was located on Riverside Drive. (Photo courtesy of Chattanooga-Hamilton County Bicentennial Library)

A location on Riverside Drive was selected, although apparently not everyone in Chattanooga was as convinced that the site was ideally suited for the manufacture of tufting machines. Lewis recalled a particular instance: "One day we were looking over the property before we started the building. A real estate agent called and said she wanted to see it. She probably had someone who wanted to buy the property.

"But when the agent asked about selling the land, my uncle said he didn't want to sell it. 'You don't need to locate a business like this along the river,' she argued. 'Yes, we do,' Uncle Joe replied. 'We might get out there and if business goes bad, we might want to go fishing.' The real estate agent didn't say anything else – she just left in a huff."

Only months after work was underway to construct the new facility and double the company's useable floor space, Joe suffered a crippling stroke while on a trip to Miami, Florida. The stroke rendered him unconscious for six weeks and required convalescent care for about a year, leaving Lewis fully responsible for the remaining construction.

"Uncle Joe wound up marrying one of the nurses who had tended to him around the clock. After the stroke, he never regained good health and remained somewhat of an invalid. But he was still good-natured, even then," Lewis remarked.

Because of Joe's insistence on encouraging self-reliance in his nephew, the health setback had no adverse impact on Lewis's business decisions. He had long since grown accustomed to operating autonomously from Joe. "But it did affect me emotionally," Lewis noted. "Joe was not only a working colleague; he was a good, close friend."

To Sell or Lease?

One of Lewis's shrewdest business decisions involved a risky but progressive marketing strategy. In the earliest days of tufted carpet, eager entrepreneurs were arriving on the scene, bursting with energy to get started in the industry – but not as well-supplied with the finances necessary to make it happen. To accommodate these would-be carpet moguls – and to stimulate sales of his machines – Lewis developed an installment selling plan.

He would ship machines on a time-payment purchase basis, often with Cobble Brothers serving as the financial institution underwriting the transactions. He also devised a leasing plan that was novel for the time, establishing a minimum fee with the rate to be increased according to actual yardage produced by the machines.

"Leasing imposed strenuous financial demands on Cobble's resources," writes Patton in *Carpet Capital,* "but it became Card's major marketing strategy to offer tufting equipment to customers who had more ambition than capital. In the 1950s and 1960s, that category would include most of the entrepreneurs entering the field."

This lease concept especially appealed to Lewis after having achieved a major advance in the development of the controlled needle bedspread machines. "I went to a customer and got an agreement for him to pay us on a yardage basis. We were leasing the machine, rather than selling it. It was my opinion, and the customer agreed, that this innovation could save him enough yarn to pay for the cost of the machine. A few months later, however, the idea was copied, so we went back to the original customer and redid the deal to sell him the machine."

While the leasing strategy was a departure from usual business practices, and it did create an added financial burden for Cobble Brothers, Lewis did not consider the approach particularly precarious.

"We generally got 25 percent of the price for equipment up front. There was not much risk, because if we had to take a machine back, we could still make money by reselling it. There were a few cases where we suffered a loss with that approach. One or two companies went bankrupt and our equipment was caught up in the bankruptcy proceedings, but that was rare."

Jerry Hendricks, who first began working with Lewis in 1958, commented on the leasing approach from his salesman's perspective. "During the time we leased tufting machines rather than selling them, the owners of some mills paid several times the price of the machines – although their return also was many times over what they paid, even on a leasing basis. They would often joke, 'Why don't you just give me the machine? I have paid for it several times already.'

"Some of those mill owners did not fully understand the potential output of the machines, but Lewis understood it quite well."

Although he really had no choice but to abandon the practice, Lewis often would think back wistfully on the brief time when leasing was their primary sales and marketing tool. "I think the biggest mistake I ever made was not leasing all of our machines, but competition made it necessary for us to continue selling the machines. It was like with IBM. There was a time when you couldn't buy machines from them. They simply leased them to you. But the climate changed when competition came along."

Maintaining a Narrow Focus

Another important decision that Lewis and Roy made together in the early years of their business was not to get into the manufacture of finishing machines, even though those had become an integral part of the carpet-making process. Once carpet came off the tufting machines, it then would be run through finishing machines that applied the secondary backing, along with doing any necessary trim work.

"Early on, we decided to concentrate only on tufting machines," Lewis said. "Finishing machinery was never part of anything we did, except for the investment. Our first love was always tufting machines. We looked at finishing machines at one time, and even talked to people about (getting involved in) the finishing business, but that is as far as we ever got.

"That's a different business," he pointed out in explaining their resolve not to diversify. "Some people don't operate or work well in both types of business. We did consider doing both tufting and finishing machines at one time, but we finally decided they were like apples and oranges – entirely different.

"It's the same industry, but separate. It could involve some of the same people, but you really needed people who knew the business to start with."

Even in later years, when Cobble Brothers began to expand by acquiring other companies, the determination not to combine the separate facets of the business was maintained. "After Cobble Brothers, when we formed Tuftco, we had a separate finishing machine business. Grover Gowin of Gowin Machinery Company joined us and we formed Gowin-Card. But we maintained separate manufacturing facilities; the businesses were so different."

Harold North, another longtime veteran of the tufting industry, commended the Cards' wisdom not only for giving their undivided attention to tufting machines, but also for being selective about which tufting products their machines would be designed to make.

When North got his start in the industry, machines for bath mats, scatter rugs and bedspreads were still being manufactured. "But Lewis had the wisdom to see that would not remain the guts of the tufting machine industry in the future," he observed. "Cobble had a virtual lock on the bath mat, rug and bedspread business, but we more or less gave it to them (competitors) after 1965 so we could concentrate on carpet tufting machines.

"Lewis and Roy focused on carpet, which I think was a combination of foresight, listening and observation. They were always trying something different. In the '60s, we made tufting machines for fleece lining and sold a lot of those machines, particularly to Japanese companies. We even made blanket machines, but doing that never proved very economical. The Cards arrived at a key conclusion: Tufting is not everything to everybody, cost-wise."

Putting Forth the Necessary Effort

As Cobble Brothers prospered, riding the momentum of America's soaring fascination with carpet, the Card brothers could have begun resting on their laurels, already assured of their legacy in the industry. However, rest and relaxation were never part of their equation – personally or professionally.

"It was all hard work. Sometimes I would get night sweats worrying about problems," Lewis said. "There was always something happening. Someone who worked for you would become drunk and get into trouble; you would have to drive to Georgia to get them out of jail. You had no choice – they were good employees, trained in tufting. You couldn't just go out on the street and find those kinds of people.

"For us the day didn't cease when we left the office. The customer would find you at home. It was a 24-hours-a-day, seven-days-a-week business."

His brother had a similar take on the expected work ethic of that time: "As my son-in-law, Charlie Monroe, once said about me, 'Roy went into semi-retirement when he stopped coming in on Saturday.' I went 30 years before I stopped thinking and believing that if you didn't work on a Saturday, at least until noon, something was wrong. It was just the routine, we thought. Everybody else did it in their businesses, and we thought that was what we were supposed to do."

Today the corporate world is governed by long-range plans or at least by "rolling plans" that are adjusted on an annual basis. That was hardly the case, Lewis points out, during the early years of developing tufting machines. Survival and the necessity of addressing challenges that arose day after day were the stark reality of the business in the late '30s, and '40s, and continued that way well into the 1960s.

And as their remarkable careers have proved in retrospect, the combination of hard work, ingenuity, timing – and perhaps a little luck – would soon be paying great dividends for everyone involved in the burgeoning carpet revolution.

New backings and new fibers would open the way for greater advances in technology. Those advances, in turn, would be applied full-bore by carpet manufacturers to provide the new American social class – homeowners – with floor coverings beyond their grandest dreams.

6 THE REVOLUTION GAINS MOMENTUM

For about 20 years, from 1937 to 1957, the winds of change swirling around Cobble Brothers and the tufting industry merely seemed like a steady breeze as innovations came slowly. Over the next dozen years, however, those comfortable "breezes" would intensify to gale force strength on many fronts. Scientific discoveries would bring about dramatic improvements in both yarns and backings. The carpet phenomenon would attract newcomers to the industry, including major corporations. And advancements in the capabilities of the machines themselves would further stimulate consumer fervor for new styles and designs.

For Lewis and Roy Card, developments would include seeing the familiar confines of Cobble Brothers itself undergo radical change, eventually leading them to pursue their craft and explore new technological horizons in other settings. Over those next 12 years a remarkable "family tree" would begin growing as machine company mergers took place and new corporate investors made their appearance. Names like Super Tufter, Singer, Southern Machine and Tuftco would surface and figure prominently in different ways, most of them bearing the indelible stamp of the Card brothers' influence.

This ongoing surge of momentum would transform the industry forever – and as Lewis and Roy were quick to realize, that was a good thing. A very good thing.

Commercial Viability – Mission Accomplished

Through the mid-50s, it was risk-takers – not the faint of heart – that demonstrated a strong commitment to carpet development. The quality of tufted carpet had not yet proved to be a fitting rival for its woven counterpart, and as we have already seen, some "experts" were quite outspoken in their skepticism about the future of tufting.

However, by 1957 the product had greatly improved and was no longer suspect. This brought about a startling shift in the business climate. Ongoing refinements in the tufting process had culminated in carpet finally being accepted as "commercially viable," and it had already surpassed weaving as the primary means for making carpet. Technology, science and commerce intersected to bolster the confidence of the new-breed tufted carpet manufacturers, encouraging experimentation with new materials, particularly far more economical synthetic fibers. To repeat the Carpet and Rug Institute web site's assessment of this pivotal period, "it was as if someone had opened a magic trunk…"

It was a virtual "chicken-and-the-egg" effect – which advances bore greatest importance for the growth of the industry? New backing materials and the revolutionary new fibers greatly reduced the raw expense of the fabric, while ever-faster, more efficient machines enabled the production of more carpet with fewer workers and less cost. Without question, the more durable fibers made the product far more enticing for consumers. But the higher production speeds and enhanced efficiencies made it possible to make appealing, functional carpet that was also extremely affordable for the first time. So who could say which was most important? In reality, everything was.

Years later, statistics would confirm the economy of carpet. An analysis conducted by DuPont, one of the major producers of synthetic fibers, showed that while the real price of consumer products overall increased 229 percent between 1967 and 1991, the average cost of carpet rose only 80 percent during the same period. Cheaper materials, plus faster production, plus reduced labor expense equaled lower cost – an ideal equation for both manufacturers and consumers alike.

The affordability of carpet even launched it into a new market: The 1960s marked the start of producing formed carpet for automobiles, making automotive carpet a totally new branch of the industry.

On the scientific front, the high cost of wool and limitations of cotton created a market ripe for alternative materials. Cotton, although beautiful and washable, had proved far less durable than desired for heavy traffic. It could be washed, but became matted and soiled quickly. The need to find equally eye-appealing but more practical carpet yarns ushered in synthetics. Nylon was introduced in 1947 and for a time dominated the market. Polyester was introduced in 1965, followed by polypropylene (olefin).

From a raw material standpoint, however, the greatest advance came with the development of new nylon yarns, which DuPont called Type 501

Nylon. This consisted of BFN (bulked filament nylon) and BCF (bulked continuous filament). Benefits of these heat-set yarns included a luxurious quality and appearance, minimizing the pilling, fuzzing and shedding that plagued previous fibers. They featured durability that rivaled wool, at a much lower cost, which enabled manufacturers to provide consumers with better products that were far more affordable.

This union of tufting machines with BCF nylon solidified tufted carpet's already promising future.

"Synthetic yarns were the best thing that ever happened to us," Lewis observed. "We originally had problems with some synthetic fibers, but it soon became evident that synthetics were easier to tuft because of the quality and consistency of the yarn."

Backings Also Go Synthetic

A similar transition took place for backing materials that served as the base into which tufting machines stitched the carpet fibers. Cotton duck backing, which was subject to as much as 20 percent shrinkage, was first replaced by jute that was used extensively between 1965 and 1975.

Jute backing, derived from a plant grown in Southeast Asia – primarily Indonesia and Bangladesh – actually came from the same botanical family as marijuana. The problem with jute was that over time it would deteriorate, and when it became wet would exude a disagreeable odor. If a bad spot developed in the middle of a 12-foot wide piece of carpet, the section had to be replaced with a sewn-in patch, not an optimal solution. In addition, since production of jute was directly related to weather, its supply and delivery were always unpredictable.

Eventually, synthetic backing was introduced to capture the nylon fiber tufts. This development eliminated both the deterioration and odor problems of the backings previously available. The use of synthetics also eliminated the shrinkage factor, resulting in considerable savings to the carpet mills. The design of tufting machines was not affected, but carpet manufacturers no longer had to factor in extra material to compensate for shrinkage. As a result, considerably less yarn and backing material were required for producing the same size carpet.

Yet another significant development was the use of tubing to guide the yarn from creels (metal racks that held the cones around which the yarn was wound) to the tufting machine. This was important because as the machines increased speed, it became necessary to deliver yarn to each individual needle with ever greater reliability.

"The technique for threading the yarn by forced air was originated in the tufting industry and adapted by hosiery mills many years later," according to Lewis. "The first creel for the bedspread machine did not use tubing as guides for the yarn. But by the time we got to tufting carpet, everything had long been tubed."

The Quest for Fast, Efficient and Better

Lewis and Roy, along with the ever-changing ensemble of carpet manufacturers, were very interested observers as these new developments for carpet fibers and backings evolved. The continual quest to create better, faster, more efficient, more versatile tufting machines remained the Card brothers' primary focus. Who would buy their machines and what the machines would produce was of secondary importance.

Although it was not their stated objective, their cumulative efforts also resulted in a product that proved to be a remarkable investment for their customers. That was one of Harold North's strongest selling points: "The tufting machine has probably the highest, fastest return on investment of any textile machine made, so investing in a new one, featuring all the newest technology, is not a risk. We always found the ROI was very high, with a ready market out there waiting to be sold."

In fact, he said, it may be that the Cards underestimated the ultimate value of their own product. "Lewis always said a tufting machine is cheap based on its return on investment. He often thought his biggest mistake was that he started out with a price for the machines that was too cheap. That made it very difficult, even impossible, to bring the price up to where it should be."

The key was continual innovation, always stretching the boundaries of imagination and understanding of the machine's capabilities. So the decades of the 1950s, '60 and '70s were marked by the ongoing quest to do more, performing feats with carpet which as recently as the 1940s could only be accomplished by hand or with slow, laborious weaving looms.

In *Bedspreads to Broadloom,* Thomas Deaton notes, "Cobble must be credited with many significant contributions in the field. It was the first to use frictionless bearings, introduce the cluster type creel, the tufting machine yarn feed, the wave-line attachment, self-setting hooks, the roller-type pattern attachment, the Universal and Scroll attachments, the Controlled Needle principle and many other refinements."

"Foremost among the inventions in the 1950s," Deaton writes, "were the pattern attachment devices. On a tufting machine this device permitted multiple pile heights, and geometric and other patterns to be made in the same piece of carpet. Previously, this was only possible in woven carpet manufacture. In 1952, a roll pattern attachment was developed and in 1955, they introduced scroll attachments."

The Universal Pattern Attachment, used in loop pile machines, controlled the yarn feed to the needle, making high and low stitches possible so that high-low nap carpet could be produced. It served as a good example of how unprecedented these innovations actually were.

"On the universal pattern attachment we developed for high-low nap, the patent office said we couldn't do that, that we could not feed enough yarn and we would break it," Lewis recalled. Naps always form the same height, he explained, but shorter ones are just pulled back to create the high-low effect. "But we are already applying the device that we were seeking to have patented and it was working beautifully. We like to have never convinced the patent office examiners that we really could do it!"

This yarn control device became Lewis's initial patent – the first of many. The device resolved a nagging problem of irregularities in loop pile fabric that resulted from the yarn delivery system. Using a differential yarn feed, yarn could be fed to the needles adequately to form even loops. This concept also led to the creation of high-low loop pile, designing a mechanism to intentionally supply yarn to the needles at two different rates as dictated by a series of clutches on the yarn feed rolls.

Products began to emerge from tufting machines which previously could only be made on looms or by hand. The first designs were limited only to geometric patterns with right angles, but over time the clutch-controlled differential yarn feed was adapted for more complex tufting pattern attachments.

Stop the Stopmarks!

Another persistent problem the Cards eventually solved was that of seam-like stop marks that ran the full width of the carpet – an irregular high or low line that resulted when the machine was stopped to repair a break in the yarn.

As Max Beasley recounted in his industry memoirs, "Every time a yarn broke, the machine would have to be stopped while the yarn was rethreaded. All gear sets have some lost motion, and this is called 'backlash.' The backlash in the gears driving the cloth feed caused a stop of this type to make a mark that is called a 'stop mark.' The first stitch after restarting would usually be a little bit shorter, then the next stitch would be a little longer to compensate as the cloth drive caught up with itself again. The result was a mark that stood out like a sore thumb.

"All sorts of brakes, springs and other things were tried to overcome this problem. These trials were done by the machine makers, as well as by all of our customers. Everybody was trying to solve the problem. It threatened to stop carpet production before it ever got off the ground. Although carpet could be tufted perhaps 20 times faster than it could be woven, all this would be for nothing if this problem could not be solved…"

The problem was resolved by having two independent gear reducers, one to drive the front roll and one to drive the rear roll. The elimination of backlash helped to prevent the release of the backing cloth while it was under tension. There could be no movement of the cloth through the machine in either direction, Beasley explained, unless the motor was actually in operation.

"This seemed to work, but it still remained to design a cloth feed transmission which was actually built for the purpose. We made the front and back rolls with spikes in them so there would be absolutely no chance for any slippage of the cloth to occur. Next, I used a worm drive with a high ratio reduction so there would be no chance of it driving itself in reverse due to the tension of the cloth… These two drives were then enclosed in a cast housing and they were connected with a variable speed belt-and-pulley drive to control the amount of tension.

"I had spent literally months trying to explain this to Spencer Michael, our patent attorney, and he had pointed out several similar mechanisms already patented, but they were used in making such things as plywood and had nothing to do with using the tension between front and back feed rolls to eliminate backlash… When he finally understood it, he got the patent in nothing flat and we had solved the problem. We then sold these transmissions to be mounted in the field on everybody's tufting machine and allowed our competitors to use it under license."

Other Key Advancements in Tufting

As noted earlier, advancements like these did not materialize instantly. They were often in response to a customer asking how to achieve a carpet effect on a tufting machine that until that time either could not be done at all or could only be done by the weaving process. Innovations would come about as the solution to frustrating production problems. And sometimes they simply were the implementation of an idea for producing a novel effect on carpet through tufting.

New synthetic materials for fibers and backings represented major strides making it possible for more costly and less durable materials like cotton, wool and jute to be abandoned. In addition to developments such as control of the yarn feed and control of the feeding mechanism for the backing cloth, numerous other innovations combined to enhance tufting capabilities and lead to the evolution of a tufted carpet that was – to the untrained eye – hardly distinguishable from its woven counterpart. Lewis and Roy Card and their companies were directly involved in engineering some of these, while some were developed by others in the industry. It would be pointless even to attempt to single out each tufting milestone, but these are some of the signature events that helped to propel carpet to ever greater heights of quality, functionality and beauty:

- *Yarn feed pattern attachment* (developed in early 1950s). This device enabled the yarn feed to be varied on command from one speed to another, changing the amount of yarn fed into each stitch to create high and low loop patterns.
- *Tube bank* (patent granted in 1958). Lewis Card invented this device as part of a scroll pattern attachment for the tufting machine. The tube bank carries yarn from one patterning roll to different points on the machine resulting in a carpet with pattern repeats across its width. This device today is still widely used to distribute the yarn on most pattern machines.
- *Shifting needle bar* (patent granted in 1962). This enabled the manufacturer to enhance the appearance of the carpet by inserting different yarns in tuft rows by laterally shifting the needle bar as the backing advanced through the machine.
- *Loop-cut pile* (patent granted in 1963). This could be considered Roy Card's most notable innovation – attaching a clip to the cut-pile looper and then, by varying the amount of yarn being fed, enabling the formation of either a high cut tuft or a low loop tuft on the same row of stitches.
- *Precision gauge parts* (developed in early 1980s). Advances in the accuracy and integrity of manufacturing these parts were critical. Moving from tolerances of .050 inches and greater was considered monumental; continued improvements have led to total accumulative error over the full length of a needle bar of .005 inches. Such accuracy became essential as machine speeds increased, with some machines approaching or exceeding 2,000 rpm, and gauges became finer and finer.

Finding the Right People

In addition to his creativity and mechanical savvy, another strength Lewis Card often exhibited was his deft ability to identify and recruit the right people to fill the right slots in the company. Harold North (who passed away in 2009), one of the most senior and accomplished of tufting machine salesmen who eventually became regarded as an icon in the industry, was a classic example.

North's entrance into the business does not seem extraordinary at first glance. He started working with Lewis in October 1957, at the age of 31, at Cobble Brothers. When Cobble was sold in 1960 and became Singer-Cobble, he became vice president of marketing, and proceeded to join Lewis at successive companies with which he was involved. But the story of how they first got together is worth noting.

North had been a menswear salesman, with Lewis his best customer. While attending the University of Tennessee in Knoxville, North had started selling men's clothing and a friend asked him to join in opening a men's store in Chattanooga. So he transferred from UTK to the University of Chattanooga and started Hixson's Men's Wear, where he became acquainted with Lewis. North worked there as the manager for about 10 years.

In the retail trade, Saturday was always the biggest day for sales, but on occasion North would slip off from work to attend a University of Tennessee football game. In October 1957, the Volunteers had an especially important game in Knoxville against Auburn. North and his wife, Frances, had cultivated a friendship with Lewis and his wife, Katherine, and occasionally would run into them before, during, or after a game.

That Saturday, Lewis had just returned from a business trip to England, visiting the Cobble Bros. Ltd. plant in Blackburn, Lancashire, England.

"It was a rainy night and I had a bad headache," North recalled. "Frances and I were already seated at a restaurant, but I needed some aspirin, so I got up to go to a drugstore nearby. In the pouring rain, there were people lined up waiting to get in. Lewis and his wife were in the line, so I invited them to sit with us at our table. His wife joined mine, while Lewis went with me to the drugstore. On the way, Lewis told me about his recent trip and I jokingly said, 'You should have me as your salesman.' 'You're right,' Lewis responded.

"Before the evening was over, I asked Lewis, 'Are you serious?' He answered, 'Sure, why don't you and Frances come over Sunday night and we'll discuss it.' And we did. Everything was exactly like we had talked about on Saturday night. So I gave my employer at the clothing store two weeks notice, and two weeks later I was on the team at Cobble Brothers. That's the way Lewis is – he makes quick decisions, and most of them are right."

The transition from clothing to selling machinery required North to devote "many long hours reading, trying to understand what a tufting machine was, and also to understand the industry.

"When I walked in on that first Monday morning, one machine was on the floor – a 216-inch loop pile machine that was going to Croft Carpet Mills in Fort Oglethorpe, Ga., owned by Easton Croft, one of the pioneers in the tufted broadloom carpet business. Back then, with jute and cotton duck backing, to finish a 15-foot carpet, you had to tuft over 16 feet because it shrunk.

"Lewis had told me, 'This industry is right on the verge of exploding, and we're going to be ready for it.' That was one of the reasons he hired me. Within just a few years, nearly all the major woven carpet mills in the Northeast started adding tufting machinery alongside their (power) looms. Lewis was right. The industry did explode, and we became busy, busy."

Not a Figment of Anyone's Imagination

Larry Gable, who worked for Card companies for more than 14 years, echoed North's sentiments about Lewis's perceptiveness about people. When he first became acquainted with Lewis and Roy, Gable was in charge of the machine shop at Bell Industries in Dalton, which was making bedspreads and later expanded into the carpet industry.

Gable quipped that he knew Lewis was the head of Cobble Brothers, but since he had worked primarily with Roy early on and had never seen Lewis, wondered if Lewis might just have been the figment of someone's imagination. Then one day Lewis came to Bell with Edgar Pickering, one of his managers, to talk with Gable about a machine that he was building.

"As they were leaving, Lewis commented to me, 'I hear that we think the same way,' referring to his understanding that I was considering leaving Bell and starting my own machine shop." Gable said. "Lewis invited me to come to Chattanooga and talk about how we might be able to help one another."

That brief conversation resulted in a relationship that has spanned more than four decades, he noted. "Lewis was a great mentor, gentleman and friend to me over the years. He was a very sharp engineer, mechanic and businessman."

Describing Lewis as a classic "behind the scenes" type of person, Gable remembered going back to Bell Industries months after he had left the company, which had been struggling for some time. Bell was liquidating its equipment and Lewis instructed Gable to do the bidding on the machinery. During the auction they would communicate from separate sides of the room.

"It was uncanny how it worked out, with no preparation. When they would come to a machine we wanted to bid on, I would do the bidding and as we got to the range of price we had agreed upon, I would move around and look for Lewis. With a very discreet nod or shake of his head, he would let me know whether to proceed with the bidding. From that experience I came up with the saying that I would rather have Lewis's *nod* than a lot of other folks' *note* when doing a business transaction."

One of Lewis's greatest gifts, according to Gable, was being able to find the right person to do a job and then supporting that individual to the fullest. "He also had an uncanny ability to develop equipment to help the industry advance."

Who Needs Competition?

Unlike manufacturers of automotive and electrical products, and even carpet, all of which faced considerable competition and continually engaged in a survival of the fittest during the 1950s and '60s, the fraternity of tufting machine manufacturers always remained small. Their number could generally be counted on the fingers of one hand. This was attributable to two primary factors: very specialized expertise and low sales volume.

"There are not billions of dollars in business to be had, so it has never attracted larger companies," Lewis explained. "And because of the limited volume of our product, not many companies had an interest in manufacturing tufting machines.

"Actually, I got in on the wrong end of the industry – carpet manufacturing is a multi-billion business. The cost of a piece of our equipment, compared to the yardage of carpet the machine puts out, amounts to peanuts. But then again, if developing and innovating the tufting process had not been our focus, there would not be a carpet business as we know it today."

Tufted bedspreads, the original impetus for development of the tufting machine, began to lose favor as consumer preference turned toward different types of bed coverings. However, enthusiasm for carpet more than filled that void. As Randall Patton pointed out in *Carpet Capital,* "By 1958 tufted carpeting had replaced woven carpeting as the consumer's choice, owing in part to a marketing blitz put on by DuPont and Barwick Mills...

"The question was no longer whether tufting would supplant weaving as the dominant mode of carpet production but who would dominate the tufting industry. The old, established woven (manufacturing) firms – rightly organized and better financed – moved into tufting and had many advantages. They were challenged by a growing number of southern firms – loosely organized, poorly capitalized, facing potential labor unrest – that had the advantages of closer connections with and greater confidence in the new technology and no outdated plants and equipment to liquidate...

"The small tufting concerns based in Georgia 'prodded the giants in the field into competition,' *Business Week* announced in June 1956. The tufting process had 'created a man-sized revolution in the hard-pressed carpet industry.' The old woven giants now 'consider(ed) tufting an integral part of their business.'"

Despite the heated rivalry among existing and emerging carpet mills, competition among tufting machine manufacturers remained confined to an exclusive handful of companies, as remains the case today. The carpet tufting machine industry is a notable exception to the fact most textile machine manufacturers have moved their operations outside the United States. Other machines are smaller in scale and much cheaper to produce and operate outside the U.S. Profits for that equipment are generated largely on the basis of high sales volumes.

Since production costs are much higher – while the volume in terms of units produced is much lower (typically less than 100 during a company's calendar year) – carpet tufting machines offer little economic incentive to produce outside of the region where the primary expertise was developed and remains.

Singer and Tufting Machines – Not a Good Fit

The Singer Company did not take this into account when it purchased Cobble Brothers in 1960, Harold North pointed out. "In the 1970's, with the growing popularity of tufted carpet, companies were eager to buy carpet mills. For instance, CBS bought Coronet, and a huge paper mill bought Trend. Big corporations bought into the carpet industry, but none of them seemed successful and there was a tendency to close mills within a few years because of lower-than-expected profits.

"It was the same with Singer when they bought Cobble. In terms of units being built, we were a very small company compared to the way Singer was making sewing machines. We worked on the basis of lower product volume (our tufting machines), but our profits (per machine or attachment) were much higher than what they were accustomed to.

"Singer eventually sold out Cobble, to Spencer Wright. But in the 1960s, it seemed Singer had more money than God, and we hardly had any competition. Their idea was to diversify – they were concerned about possibilities of future competition from Italy and Japan. They were buying other companies, in terms of size, like ours – such as Fidelity Machine Company, Supreme Knitting Machine Company and others. But later it became a disaster. Singer did not realize the total picture. When they bought us, we just didn't fit Singer's business culture and structure."

Lewis confirmed the view of Singer's miscalculations about the industry. "They would send their 'experts' down every now and then, but we all were trying to learn the business, so there were no real experts."

Prior to Singer's arrival on the scene, the issue of competition within the tufting machine industry had become inconsequential. In 1959, Cobble Brothers acquired Super Tufter Machine Company, the formidable competitor that Albert Cobble had founded after parting ways with his brother, along with Southern Machine and BTM in England, acquired through its subsidiary, Cobble Brothers Machine Co., Ltd.

Reflecting on Those Early Days

Jerry Hendricks began his career in the tufting industry with Super Tufter prior to its acquisition by Cobble Brothers and vividly recalled what those early days were like.

"I went to work there when it was located in an old Army barracks in Fort Oglethorpe, Georgia. One section served as the site for the machine shop, while another area was designated for the assembly department. I was initially assigned to PARCO, run by Jim Pardue, who worked for Super Tufter making loopers and mending guns. The first couple of weeks I worked in the stock room; my father, James, was a machine shop foreman."

Super Tufter executives and employees in the early 1950s.

Hendricks's "honeymoon" in his new profession was short-lived. "Super Tufter was a union shop, and soon after I got there we went out on strike and stayed out for several months. During the strike my job was to take people back and forth on the picket line. We even picketed in front of Cobble Brothers, which was not a union shop.

"After the strike ended, there was no way that I could recapture the earnings I had lost. I thought, 'What a way to start my working career – striking, instead of trying to learn something and understand the business.' Ironically, after being on strike for months, we all went back to work for virtually the same money."

Industrial health and safety regulations were not nearly as rigid or comprehensive during that time, either. "In Super Tufter's case hardening department, Vernon Mason ran the welding shop, doing fabrication and heat treating. A large cyanide pot was used in the heat treating process, and it was not unusual to have it in use during the day. Everyone was warned when it was in operation because it gave off toxic fumes."

Participating in a lengthy work stoppage and being diligent to avoid toxic fumes – not the ideal launch of a working career, but Hendricks hung on to become a mainstay of the industry over five decades.

During his early days at Super Tufter, Cobble Brothers had already started developing a scroll machine – a pattern machine that could create free-flowing designs on the carpet. The geometric or "eccentric" roll pattern attachment used was purely mechanical.

"As the scroll pattern idea was being developed, a group at Super Tufter put in very long hours to develop our own version of it. One of my responsibilities as a new employee was to be a 'go-fer,' and one night Bud Cobble, Albert's son, told me to go to the Krystal restaurant in Fort Oglethorpe and get hamburgers for all the people who were working. There were about 10 guys working that night. He gave me a $100 bill, and I thought, 'You can get a lot of hamburgers for $100!' In those days, Krystal sandwiches cost 10 cents each.

"The guy at Krystal could not believe I wanted so many hamburgers. The order took up all five of their hamburger grills, and even at that, I went back to work with $35-$40 worth of hamburgers. My station wagon was filled with those Krystal hamburgers! The men were very appreciative, and each of them ate 8-10 hamburgers, but we still had hamburgers left over for several days."

Even in those days when long-term relationships with employers were common, Hendricks learned a memorable lesson about how tenuous a job can be.

"We had an intercom system between our two plants," Hendricks recalled. "The intercom had a three-position switch: 'off,' 'talk' and 'listen.' One day I and a young guy named Jack Schroeder, an assistant in the stock room, were there and heard about a meeting to discuss layoffs that everyone knew were coming. Jack, the nephew of one of the foremen, and I went to one of the offices and flipped the intercom button so we could listen in on the meeting. We heard someone say he was going to read the layoff list they needed to review.

"The first name that he read off was Jack's. Talk about seeing someone's expression change dramatically! He quickly lost interest in hearing what else was being said."

One key to retaining a job and advancing within a company, Hendricks noted, was versatility. "In those days, like most people, I wore a lot of different hats. I started out in the tool room, handing out tools – drills, reamers, taps – to the machinists. But if something was needed – sweeping, a delivery – younger guys who worked in our area would be called on to do it.

"Working in the stock room was educational for me, because I had to learn the rhyme and reason behind each tool. I gained an appreciation for small drills, larger drills, and reamers because the precise tool was needed to remain within the tolerances of the drawing specifications for holes being drilled. Guys coming through the machine shop had to know, or learn, how to run about every machine – a saw, lathe, or milling machine. You pretty much had to do it all.

"They were referred to as machinists, but today jobs are more clearly defined, more specialized. Of course, this is also a reflection of the complexity of machines today. Back then, the most advanced machines only cost $75,000-$80,000. Today, some of the high-tech, computer-controlled machines may cost as much as several million dollars. While things may be more sophisticated now, the basic techniques are still used today; mechanical laws still must be followed."

Discovering the New World – of Electronics

Hendricks noted that work on the scroll machine Cobble and Super Tufter were developing simultaneously signaled tufting's initial attempt to combine electronic and mechanical capabilities.

"When we were developing the scroll pattern machine, it was the first step into the electronic era for tufting machines. We had tried it a bit before, but that really got us started and moved us into an electronic mindset. We engineered machines that used a hard paper tube about 20 inches long. A machine was used to cut grooves in the tube and a pattern was created by what was cut into the tube. Later we used a phenolic drum and an electronic finger that sent a signal to a solenoid, causing a needle to sew or not sew, or vary yarn feed rates (fast for high pile carpet, slow for a low pile carpet).

"Later we went to a brass, 12-inch cylinder that had a design or pattern cut into it. Similar to the paper tube, the pattern was used to send electronic signals or impulses to cause the needle to go low or high.

"Then, in the early '60s, we employed photoelectric technology, going from a brass cylinder to clear plastic, with a lamp inside a plastic drum. Outside of the drum, an acetate film was used with a specific pattern drawn on it. In the design room they would take a sheet of plastic, lay a paper design on it, and with a sharp knife or razor, cut the design into the covering on the acetate film. They would drape this film around the clear drum, and as the drum passed over the bar of photo cells, if the light came through then the yarn at the needle would go high. If there was no light, it would go low.

"In the years since, there have been a lot of improvements on the gear we used. Eventually everything evolved into solid-state equipment. This is what we use today, only it is now all computer-controlled, rather than triggered by electronic impulses."

When Super Tufter was taken over by Cobble, a conscious effort was made to combine the best of what both companies had to offer. "Cobble was making standard loop pile and cut pile machines of any width up to 216 inches. When Super Tufter became a part of Cobble, a decision was made to create a 'family machine,' taking the best features of both the Super Tufter and Cobble machines," Hendricks stated.

"This family machine let us focus on improving tufting in many respects. For instance, Super Tufter had a dial-type mechanism to adjust the bed rail and adjust the needle strokes. By combining that with the superior efficiency of the Cobble machine, we could merge the ease and accessibility of changing settings with greater efficiency and accuracy.

"When I went there, we were making cut pile bedspread machines. We developed a lighter frame to operate at faster speeds. We also made rug machines, before eventually making the shift to carpet machines."

Southern Machine, Followed by Singer

Less than a year later, Cobble Brothers made another significant acquisition. The 1960 purchase of Southern Machine Company, a smaller manufacturer that had been established by former Cobble employees, enabled Lewis to fashion a nearly perfect environment – a market with virtually no competition.

That soon changed, however, when Singer purchased Cobble Brothers. "Singer was a huge company in those days, with factories all around the world, including Russia," Lewis pointed out. "Singer was in an expansion and diversification mode, buying all kinds of textile machinery companies, including Cobble Brothers."

Because of Singer's overall corporate size, it became necessary to divest one of its tufting machine entities to avoid accusations of trying to monopolize the industry. Southern Machine, its smallest division, was the choice. Because there were no other prospective buyers, it was sold to some familiar faces – Joe and Albert Cobble. Years later, that sale would provide leverage for establishing a brand new tufting machine enterprise that would restore competition to the industry in a significant way.

An aerial view of Singer-Cobble after it had relocated to Riverside Drive in Chattanooga. (Photo courtesy of Chattanooga-Hamilton County Bicentennial Library)

North noted the dramatic change in the company culture after Singer assumed control of Cobble Brothers and how difficult it was to adjust to it. "In the early days, we were moving so fast, it was hard to envision what would be happening five years down the pike. When Singer took us over, one of the hardest jobs I ever had, one that I hated, was forecasting – projecting future sales. With Cobble, I had never had to forecast – I was too busy trying to keep up with my other responsibilities. So when Singer told me to do forecasting for the coming year, I was literally pulling figures out of the air. My forecasts were always very conservative, small enough to be sure I was going to exceed my projections.

"When Singer came in, about 10 percent of its money was spent on research and development for other types of machinery, deviating from conventional tufting machines. Much of their energy went to R&D, developing new products, but hardly any of those ideas came to any significant fruition. For a time we even got into packaging machines – experiments like that cost us a lot of money at Singer-Cobble."

Checking Out His Options

When Cobble was sold to Singer, Lewis had a three-year employment contract and a five-year non-compete agreement, so even though he also struggled with the new work atmosphere, his options were limited. He retained his management role, but was reporting to Donald Kircher, a World War II vet who had been disabled by wounds suffered in battle.

Max Beasley recalled, "(Kircher) was a very strong individual, just like Lewis. A year or so after the purchase of Cobble, Lewis told him that he would be leaving the company, since his working contract ended in 1963, and his non-compete overall finished in 1965. When Kircher asked him why he wouldn't stay, Lewis gave a famous reply: 'If I stayed, I would want your job, but you seem to be well-entrenched.'"

To his credit, even though he found himself in an unfamiliar reporting relationship, Lewis continued to give Singer a fair day's work for a fair day's pay, according to Beasley. "Lewis worked just as diligently for the Singer Company as he did when he was part owner. Although Cobble produced only about two percent of Singer sales, they contributed a much larger portion of the profits."

Eventually the time came when Lewis, ever the entrepreneur and independent innovator, needed to make a change. Longtime tufting veteran Larry Gable remembered what occurred next. "He came and talked with me, and Roy T. Card Company was being founded. He asked if I would come to Chattanooga to help in getting Roy T. Card started, and then I could return to Dalton and start my own machine shop. True to Lewis's word, that is what we did. After 18 months, I went back to Dalton."

Roy T. Card Company was created to repair and modify tufting equipment, and eventually make its own brand of tufting machines, called Card machines. Later Card & Co., Inc. combined with a finishing company in Dalton, Grover Gowin Company, the latter of which became Gowin-Card.

Singer-Cobble Gets Some True Competition

When Lewis Card severed ties with Singer-Cobble in 1965, it signified the close of a remarkable era. His uncles, Joe and Albert, had founded Cobble Brothers in 1937 and Lewis had arrived in 1939, soon beginning to make his mark – not only in the company but even more so in the emerging tufting industry. Thankfully for the industry, 26 years after arriving on the scene, his work was far from completed and his presence would continue to be felt in the world of tufting for many years to come.

Competition, which had become virtually nonexistent in tufting manufacturing for years, was about to reemerge. A new company, Tuftco, was poised to surface as the latest member of the tufting brotherhood, as we will see in the next chapter. This development some years afterward would indirectly result in the appearance of another, even more formidable player to explore the limits of tufting technology.

7 SHIFTING THE COMPETITIVE BALANCE

The entrance of Singer into the tufting machine industry had precipitated several noteworthy developments. There had been a changing of the guard, with many of the early pioneers moving toward retirement so they could reap the tangible fruits of their labors. George Muse returned to Sugar Valley, Georgia and his family's business for good, while the Cobble brothers had also stepped away, concluding their contributions to the rapidly evolving industry.

In addition, it was not long before Singer-Cobble and the Cards arrived at a figurative fork in the road, determined to follow divergent paths. One was narrow, the other increasingly broad.

Not content to concentrate on its core competency – the sewing machine – Singer had chosen to pursue a course of diversification. Its leadership looked to variety as the key to future growth and profitability.

Jerry Hendricks, having entered the tufting machine industry in 1958 with Super Tufter and moving to Singer-Cobble in 1960, noted, "In the stock room, we had lots of parts for supplying hosiery mills. If times were slow, that still kept us busy. We also made other types of machines – for instance, railroad tie doweling machines (designed to keep railroad ties from pulling apart); a packaging machine division, and finishing equipment. Singer wanted diversity in the kinds of machines we made. At one point they had 750 employees."

In contrast, Lewis and Roy refused to become distracted or to change their focus, convinced that the potential of tufting technology had barely been tapped. Lamar Card, Lewis's elder son, remembered a time that – in retrospect – revealed the sharpness of that focus.

"One day, when I was 14 or 15, Dad handed me two pieces of plastic that were stuck together. I pulled them apart, and Dad told me a way had been devised to hold things together without using tape. About 10 years later I saw the same material on the market – it was Velcro. Dad had nothing to do with it, but at one time he could have. To me, this was evidence of his determination to keep the main thing the main thing – focusing on the tufting process – or as they say in baseball, keeping your eye on the ball."

Roy, who did not have the long-term contractual commitments of his brother, parted ways with Singer in 1964. He formed Roy Card & Co., a service business for repairing and modifying tufting equipment. A primary benefit of the company was to keep him – and not long afterward, his brother – linked to the carpet industry while their obligations to refrain from competitive ventures were being fulfilled.

After leaving Singer-Cobble in 1965, Lewis did not have to travel far to find his next job. He joined Roy's company, also located on Riverside Drive across the street from Singer. Being prohibited from entering into direct competition with Singer-Cobble for two years due to his non-compete agreement, Lewis set about selling jute for carpet backing. Lewis admitted, "For me, selling jute wasn't that big a deal – it just gave me an opening to stay in touch with our customers and remain connected to the tufting business."

While sales had long been part of Lewis's job description, being a jute salesman allowed him to bide his time and served as a bridge for returning to what he truly loved to do – develop and refine tufting technology. In business with Roy, through both Roy T. Card Co. and then Lewis Card & Co., they teamed to build what they called "Card machines," but the next major step for their careers – and the industry – came in 1969. This would come about through another Cobble-Card connection.

New Player Enters the Tufting Game

Jack Frost had been involved with Southern Machine since 1963, serving as a financial advisor to Bud Cobble. He had done tax work for Albert Cobble, and through that became involved in day-to-day business matters "more than the ordinary accountant," drawing from his experience in manufacturing prior to moving into the accounting field.

Initially, plans were being formulated to take the company public, but fate stepped in. First Joe Cobble died, succumbing from an extended illness that dated back to the severe stroke he had suffered while still involved with Cobble Brothers.

At that point, in 1969, Bud Cobble, Albert's son and head of Southern Machine, and Lewis Card initiated discussions about starting a new tufting machine manufacturing company, which would be called Tuftco. Lewis had been the executor of Joe Cobble's estate, and as a result became integrally involved in these deliberations since Joe and Albert Cobble each had owned 50 percent shares in Southern.

Bud Cobble standing in front of Southern Machine
Company that he owned with his father, Albert.

"When Lewis came into the picture, he decided to bring in both Card & Co. and Gowin-Card, which was partly owned by Card & Co. I became involved with Lewis and Roy through my business relationship with Bud," Frost said.

"We did take the company public, as we intended, but Bud died of a heart attack two weeks afterward. My interest had been to take Tuftco public and then move on to other things, but when Bud passed away, Lewis and I agreed that I should come in as the president."

The loss of Bud was significant, Frost noted, because he was a seasoned veteran of the tufting business, having been an active part of it since he got out of high school at the age of 18.

"He was primarily involved in mechanics and sales, but he was also a good technician, like Roy Card," Frost noted.

Lamar Card vividly remembered the considerable impact Bud Cobble had on the industry – and people working in it – despite his departure from the scene while still relatively young. "Bud and Dad were competitors in business in the early period. Bud's father, Albert, had gone into competition with his brother, Joe, and that competition extended between Bud and Lewis, particularly since Joe was a father figure for my dad.

"Despite that, Bud and Dad were very close. My father has even included a couple of interesting shots of Bud on the wall in his garage in Chattanooga, where he has assembled a photo display.

"Personality-wise, he and Dad seemed to be the opposite number. People who didn't know them would have guessed they were total contrasts. Dad always had his nose to the grindstone, was an achiever, and kept firm schedules. Bud was a party guy, although he knew how to work, too.

"I would describe Bud as the uncle every boy wanted – fun-loving, always pushing the edge, introducing you to new experiences. At the same time he was brilliant, and I believe he was one of the more significant contributors to tufting during his brief lifetime."

A Formidable Competitor

Until that time, Cobble Brothers and its successor, Singer-Cobble, had gained a virtual stranglehold on the tufting machine market. In the early 1960s, of the estimated 1,400 machines being used, approximately 85 percent of them had been produced by Singer-Cobble. Suddenly Tuftco stepped into the fray and the level of competition began to escalate immediately. Ultimately, Singer would respond by bowing out of the tufting arena.

By the mid-1970s, Singer had revised its thinking about diversification, beginning the process of divesting certain product lines so it could fortify its concentration on the sewing machine market. In February 1977, privately held Spencer H. Wright Industries bought Singer's U.S. and overseas tufting operations that they have continued to maintain to the present.

While status quo at Tuftco would also prove to be relatively short-lived, the company's initial years were marked by an unusual esprit de corps. The Card brothers again were able to channel their energy and talents in their vocational niche and surround themselves with a gifted team of men that continued to challenge the barriers of tufting technology.

Larry Gable now works as the senior partner of Metal Crafters, Inc., which manufactures replacement parts used in tufting machines. He was employed by Roy T. Card Co., Lewis Card & Co. and Tuftco for more than 14 years, before forming his own companies in Dalton and Chatsworth, Ga. Gable recalled Tuftco's early years.

"Lewis and Roy were the daddies of the tufting industry as we know it today. I recall in the Tuftco days, we could absolutely do no wrong. We had more orders for machines than we knew what to do with. We hardly ever had night meetings, but one day Roy did decide to call an evening meeting at our plant on Holtzclaw Avenue. I vividly remember that because I was newly married and had to assure my bride that this was not going to become a regular occurrence.

"We (the Tuftco staff) went out to dinner and then returned to the plant to talk about various projects we were involved with. Somebody made some kind of comment and Roy laughingly replied, 'Where did you come up with that?' Edgar Pickering, one of our sales managers, responded, 'We all learned it from you, Granddad!'"

Working in the Card companies, according to Gable, "was always a lot of fun" – despite long hours and the intensity of keeping up with the changing demands of the industry. In many respects it was still a new industry and every day participants could feel that they were on the cusp of some exciting new development or innovation.

'Proud Moments' for Innovation

As an inventor, Gable said his "proudest moment" was at Card & Co. in the early 1970s when he was instrumental in development of the Hydrashift that enabled the needle bar to shift, moving needles from one looper to another and making possible a wider variety of designs and patterns.

He explained the needle bar had cams on both ends and drive rods that would slide the needle bar while keeping to tolerances within thousandths of an inch. "The cam worked well in the early days, but it was noisy. And when cams wore out, it took one to two hours to replace them, then more time to readjust the machine – three to four hours in all.

"In those days we had a tape-controlled milling machine – an early stage of computer technology for the carpet industry. The machine was accurate to 4/1,000 of an inch. It seemed if we could adapt that technology to the shifting needle bar, we could solve the problem.

"Wallace Hamell was in charge of research and development at Tuftco then, and I showed him how precise the milling machine was. Moog was manufacturing our milling machines in Buffalo, New York, so Wallace went to visit Moog to see what could be done. They developed a control box smaller than an eight-track tape machine at the time, enabling us to change the box when needed and push a button to restart in just 30 seconds. Wallace, an engineer at Moog and I all got our names on the patent."

It was during his time at Cobble that Roy Card achieved perhaps his most significant innovation as well. He designed a new looper mechanism that made it possible to place both cut and loop pile stitches in the same row. Implemented with existing pattern equipment, this device enabled the creation of strikingly new carved effects by placing cut and loop pile nap next to each other. Combined with different colored yarns, this cut-loop apparatus allowed the controlled burying or hiding of loop stitches below the level of adjacent cut pile stitches, resulting in a distinctive look with random tones and colors.

Roy's role in heading up the sample department clearly paid healthy dividends. Since the sample machines were constantly being converted from loop to cut pile, and back, and gauges were being changed just as often in accordance with customer demands, Roy received ample exposure to the fine points and capabilities of tufting machines.

A key element in the cut/loop innovation was a step that he termed "backrobbing." As Max Beasley describes it in his memoirs, "The height of loops in a loop pile fabric could be determined strictly by the amount of yarn fed (to the needles). Loops could be formed higher than needed, then 'backrobbed' out of each stitch successively simply by supplying less yarn. By having two yarn feeds and threading the yarn through these in groups, then having each driven with two speeds, it would be possible to make a pattern like a checkerboard by alternately changing speed on these two drives."

Succeeding By Trying, Trying, Trying Again

While the Cobbles and the Cards set the pace for many years in the tufting machine industry, their contributions certainly were not accomplished in isolation. Frost noted that he was privileged to observe and benefit from the collective expertise and inventiveness of many gifted individuals.

For instance, he said Bud Cobble and Max Beasley worked together very closely in the early days. "I hired Max into Tuftco, first as assistant to the president. He worked as a director for me, and eventually became the head of the engineering department for us.

"Over the years we had a number of good people who were inventors and very important in the development of the tufting machine. For example, we had situations where we were using coarser gauges and wanted to develop closer gauges – being able to move the needles closer together. We were working to narrow it to 1/10-gauge. Hoyt Short had been a racing boat mechanic for Bud Cobble, and Bud had brought him to Tuftco to become a tufting mechanic. Hoyt had been working on the gauge problem, and one night he finally got to the point where he could operate the machine at 1/10-gauge. He ran it all night to make sure it would continue to work.

"The next morning when I arrived for work, my office was filled with carpet that Hoyt had produced throughout the night."

Long workdays were the norm for this early stage of the tufting machine industry, according to Frost, whether for testing new refinements or planning for major corporate changes. "When we were putting Tuftco together, Bud and I would work here at our Holtzclaw Avenue plant until 10-11 o'clock at night, or we would go to Lewis's office on Riverside Drive and stay until 10 or 11." Unfortunately for Bud Cobble, he did not live long enough to experience much of the fruit of those late planning sessions.

Gifted Brothers – But Gifted Differently

Being president of Tuftco, Frost was able to observe both Lewis and Roy from a unique vantage point. He grew to appreciate the men's similarities and differences.

In terms of personality, he said, Lewis and Roy were "quite a contrast. Lewis was more intense. He was pretty well a driven type of person. Roy was the more laid back and relaxed of the two.

"Roy was a very good technician and mechanic, very knowledgeable as far as tufting machines go. He was more into the sewing and production end of the business. Lewis was that as well, but also a very savvy businessman.

"Both Lewis and Roy were very sharp mechanically, but Lewis by far was more the entrepreneur. He was a risk-taker, had a quick intellect and was a fast study – Lewis picked things up well, and really enjoyed a challenge. He was a rolling stone, with many different interests and pursuits."

By way of example, Frost noted Lewis's entrepreneurial interests were not limited to tufting machines. "Together he and I got into the skating rink business, house building and real estate development. One of my first contacts with Lewis was when he had a subdivision he was developing and had some audit work he needed done. We got into modular home building and were way ahead of our time in that.

"One day Lewis was riding around Florida, and in Sanford he saw an old Hunt's tomato canning plant that was owned by the Hunt family in Texas. The plant no longer was in use, and Lewis saw that it could work well as a mobile home plant. We arranged for a representative of H.L. Hunt to fly into Lovell Field in Chattanooga; he and I met over breakfast and we agreed to buy the plant sight unseen."

Regarding Lewis and Roy's legacy to the industry, Frost stated, "They were very significant, instrumental figures in the industry, especially in the early days. They were true pioneers, although there were a lot of people that made major contributions back then. Some received more credit than others. The industry as we know it today is the sum total of the collective contributions of many people."

Taking the Simple Approach

"Max Beasley used to say that a lot of the innovations came from Lewis simply asking questions like, 'Could we do this differently? Can we do it another way?'

"Lewis was, in a word, tenacious. If he had an idea, he would work it until he found an answer. He might show up at the plant at 4 o'clock in the morning, just to pursue his idea until he found something that worked," Frost said.

In his writings, Beasley cited a typical example that involved development of one of the key advances in the industry, the Universal Type Pattern Machine. "When we fitted it with the solenoids, they had to be quite large, and when they were seated, they made a noise like a cannon. When all of them got going, it sounded like you were in a battle zone. At this point Lewis said, 'Why don't we get some electric clutches and eliminate the solenoids?'

"Obviously this was the thing to do, and this greatly simplified the machine. Using the same basic frame, I had a common drive shaft from which I had separate drives going to each clutch. This required eight chains, with eight idlers, and filled the attachment up with parts. I thought I needed all this, but after we saw it, Lewis said, 'Why not drive each of the two speeds with only one chain which drives all the clutches? Only the ones under power will pull the rolls.' This was done immediately and the pattern machine was born."

Interactions like these helped impress upon Beasley a basic principle of innovation: "I have always thought that simplicity was the outstanding criteria of any design, next to function itself, and I had a sign on my desk for years that said, *'The first thing a weak mind does with a complicated problem is to further complicate it.'*"

More often than not, solutions to problems in designing tufting machines came from streamlining concepts, he pointed out, noting how on several occasions Lewis helped him to resolve persistent problems through "a method of simplifying my first ideas."

Tufting – An Art or a Science?

This emphasis on simplification was not a reflection that the tufting process was not complex. In fact, as machines grew more sophisticated and could be run faster, the premium on precision increased accordingly. To reach higher standards for accuracy, greater reliance was placed on empirical support for new innovations, rather than creative artistry. Beasley recalled a comment by Frost that underscored this strategic shift. "(He) always had a habit of getting us together for a staff meeting every week… We were having quite a few problems with the machines, and Jack's method was to attack one problem at a time and stay with it until it was solved…the object was to remove the 'art' and make the operation of tufting more of a science."

Roy Card took a similar approach while he was heading the technical service department, according to Jerry Hendricks. "When Roy headed technical services, he had daily morning meetings to evaluate job progress – going over with technicians and evaluating if they needed help. He would interview everyone and if he found that any problem was happening with frequency, he knew that was something he needed to work on relative to that machine."

Roy Card (right) stands with customer Jim Horner in front
of the Card & Co. offices in the late 1960s.

Allen Neely, who began as a machinist but has been involved in virtually every area of tufting machinery manufacturing, agreed on Roy's thorough approach to problem-solving.

"When I started dealing with him on a routine basis, I realized that all Roy wanted was complete information. He needed to know you had done due diligence (in researching a problem or exploring a need), and once he was confident that had been done, he was fine. But if you would come to him with incomplete information, if you didn't have all the answers, he would tear you apart with questions. So you wanted to show that you had done all the research, had asked all the pertinent questions and knew the right answers for them."

Problems did not always have a mechanical basis. Roy had a special knack for ferreting out the source of the problem, even when it could not be attributed to a machine malfunction, according to Beasley's writings.

"The problems we had to resolve were not only issues concerning our machines, but also all of the raw materials that were being developed – such as new synthetic fibers. With various kinds of synthetic backing and yarns being developed, we had to make improvements to our machines – and also learn about environmental factors that could affect production.

"For instance, when nylon is extruded, it has to be done in an atmosphere that is controlled. If there are fluctuations in the environment, it could affect the color of the yarn – darker or lighter – or create other issues, resulting in irregularities in tufting or the appearance of the carpet. There could be streaks or uneven bands that show up," Beasley wrote.

"Roy had an innate ability to separate and distinguish a machine problem from a yarn or raw material problem. Sometimes a customer might think a problem in production was machine-related, when it was not. But Roy would not stop there – he also would help the customer find out how to solve the problem even if it had to do with yarn or raw materials. I really think he had a God-given talent for being able to separate those issues and get to the cause in a hurry. In our business, we could have worked on a problem for several weeks before we realized it was not due to our machines, but he was able to cut to the chase very quickly.

"When customers would call and ask us to work on one of our machines to fix a problem they were having, Roy devised simple techniques and tests to determine the source of the problem – whether it was ours or theirs. Once we confirmed that our machine was tufting the way it should, we could then look elsewhere to identify the source of the problem."

Computers – Initial Skepticism Overcome

The introduction of computerization had an almost immediate impact in the quest for both simplicity and sophistication, although it took some time to convince many of the industry's veteran practitioners. Take longtime salesman Harold North, for instance: "I am ashamed to admit it, but when I was first asked about the application of computers to tufting machines, my thoughts were very negative. I could not see any application whatsoever.

"Computerization has brought the industry more or less from darkness to daylight. We now have the ability to electronically control pile height and stitch rates. It beats the devil out of the old mechanical way we used prior to that, using clutches to control what we needed to do. Using electronic impulses, rather than the mechanical, metal-to-metal approach, enables the machines to work faster, more efficiently and with more creativity."

North was not alone in his initial lack of appreciation for computers, particularly their viability for the tufting world. In his writings, Beasley included this admission: "I had no experience with computers at all, and when I first saw how to make a drawing with the Autocad system, in response to a question I replied that it wouldn't beat a good draftsman. It actually took me several weeks to realize some of the things the Autocad could do, (including) the work of several draftsmen!

"You will gladly do more drawing than you would have done by hand because it is so much easier to do. Take, for instance, a part which you will use over and over, such as a flange bearing, a pillow block, or a gear. With the computer, you make an accurate drawing of it one time, then you can insert it into any drawing you are making and it takes no time at all. It will be to scale, and you can move it around and look at it to your heart's content. You can make drawings of all the frequently used parts and can call them up and use them any time you need to. ... The Autocad can also duplicate a part as many times as you tell it to, flip it into a mirror image in any direction you tell it, and any number of other things which would be very time-consuming if done by hand."

Once convinced of the computer's application and value to the industry, however, there was no hesitation to put it to full use. Bruce Dyer, who has been involved in the tufting industry since 1969 and one of the Cards' most satisfied customers through the years, attested to Roy Card's eagerness to integrate new technology into his trade.

"Roy Card was a big player in getting into new technology," Dyer noted. "As new developments allowed, his machines were always custom-made and high-tech. Roy was the leader in the transition of the carpet industry, taking it from the Model T to the Mercedes Benz.

"He had the vision to go with changes and high-tech innovations – servos, computers. And now, the high-tech segment of the carpet industry is the only one that is growing, going up."

The Proof Is in the Selling

Even the best product, of course, is of little value if it never finds its way to the customer. This truth was not lost on the Card brothers, particularly Roy, who excelled at customer relations. With Cobble Brothers and Singer-Cobble, he had functioned primarily in the technical side of the business, although in the sample departments he did interface extensively with customers. At Tuftco, he was able to broaden his professional scope, becoming more personally involved in sales and demonstrating his strengths in that area as well.

By combining his technical savvy and interpersonal skills, Roy acted according to his belief that in marketing tufting machines, there was no need for glitzy, clever salesmanship.

Brad Card, Roy's younger son who is now a sales executive with CMC, recalled his father's advice. "'You may be a good salesman,' he has always said, 'but the machine has to sell itself. You can sell the first one, but after that, service and relationships are of primary importance. If the machine does not do what it's supposed to do, you're not going to keep the business.' Having a good product makes the sales job really easy.

"You hate to lose any machine orders, but if you lose one, sometimes that's the best sales tool – the customer knows what they had (from you), and compare that with what they get from the competitor. You always have to stay in front of the customer and provide quality service to them when it's needed."

Putting the customer's needs foremost was always a hallmark of Roy's sales approach, according to son Brad. "Sometimes a company has a policy stating not to ship anything to them until they have provided a purchase order. But Dad always believed in doing whatever we could to take care of the customers to keep their machines running. If they needed a part or service, we would try to provide it in any way possible."

North, looking back on 50 years of experience in the industry, pointed out that although tufting machines are complicated, it's not the complexity that closes the sale.

"In my position, you have to have some mechanical background, but mine was very limited compared to that of Lewis and Roy. I needed to know something to sell hundreds, even thousands of these machines around the world the last 30-40 years. However, when I'm speaking to the founder, owner, president or CEO of a carpet mill, 99 percent of them are not really interested in how a looper works on a tufting machine. They want to know what a machine can do for them, how much money it will make in a given period, what the payback will be – things like that. They are interested primarily in the finished product, not in how to produce it.

"There is a sales axiom about knowing your product – but not necessarily going into the guts of the operation. What it is going to do for your customer is what sells. This applies to sales in just about any business or industry."

Neely recalled several examples of the down-to-earth philosophy Roy embraced regarding sales and productivity, memorable statements he would often repeat, including:

- "If you can make a niche machine, everybody in that market has to have one."
- "Inspect what you expect."
- "The customer in the field is the No. 1 priority."

"And Roy would always say that you had to sell 10 of a given machine just to cover the development cost, not register a profit or loss," Neely recalled.

On the Selling Block Again

In a series of events with striking similarity to developments with Singer, Tuftco was sold in 1972 to a New York holding company. After several years, again uncomfortable with the constraints of absentee control, the Card brothers elected to leave Tuftco. Eventually they would become involved in developing another equipment manufacturing business, CMC, as we will see in chapter 10.

Commenting on the brothers' decision to sever ties with Tuftco, Frost said, "Even though we later got into a lawsuit over a particular patent, Lewis and I still stayed friends, even though we were competitors. Despite the fact that we are very competitive, a good relationship has remained between Tuftco and Card-Monroe (the brothers' eventual and – as it has turned out – last professional destination). Each year we pay royalties to them for some of their patents, and they pay us some royalties for our patents. Lewis and I still get together occasionally. In terms of our approach to business, he and I have been a lot alike in our thinking."

Reflecting back on the hard work and challenges that he and his companies faced during his decades in developing and refining tufting machines, Lewis commented on the importance of unwavering belief in the work they were doing.

"This has been a tough business. My Uncle Joe gave me the leeway and said, 'Go to it.' The fellows (whose companies) I bought all believed they had taken me to the cleaners. Everyone, including Uncle Albert, thought they had made a good deal at our expense. For Southern Machine, any deal was good. But these all came along because I had faith in my understanding of what we were doing – and we were spending good money to get there."

The "tough business" would become more acute in the late 1960s and early '70s as the so-called "Golden Age" of U.S. and Western economic growth slowed and came to an end. After nearly 20 years atop the crest of this widespread prosperity wave, however, the carpet industry – as well as many other industries – experienced the end of its economic boom.

By this time, all the "basics" of carpet had been put into place. Tufting, backing, secondary backing, yarn and colors had all been developed in small increments. Color and its application, shades and styles became the selling points for customers. Despite the attraction to the latest fashions in carpet, demand by consumers having to conserve their financial resources took a dramatic dip, especially since most of them already had carpet of one kind or another. One result was a separating of pretenders from the true contenders among carpet makers.

Although they were making the machines and not the carpet, companies like Tuftco and Cobble felt the pinch as well. "We had a six-month backlog (of orders) in 1973, but within a week that had shrunk to a 5-6 week backlog as customers canceled orders right and left, as a consequence of the OPEC oil embargo," Lewis recalled. "We struggled through it, but came out the other end stronger. As has often been said, you can really tell how good you are as a manager during bad times. In good times all you have to do is just keep up (with the orders)."

Troubles in Transit

Success in manufacturing tufting machines had always depended largely on a commitment to high quality and excellence. However, there was one other important detail – the logistics of delivering the machine safely to the customer. Most of the time this was not a problem, but as Harold North recalled, there were a few unfortunate incidents.

"In the latter days of Lewis Card's time with Singer-Cobble – while he was completing his working commitment to them – 5/32-gauge scroll loop pile was by far the No. 1 residential carpet machine going. We sold them to mills in Dalton, California and Canada," he stated. "Our biggest customer was Coronet, which was well on its way then to overtaking Barwick as the largest American carpet mill. During that period in the mid-60s, we had a standing order with Coronet to ship them one of our machines every month.

"In the meantime, the founder of Salem Carpet Mills came to see me at Singer-Cobble. He purchased one of our machines, also a 5/32-loop pile machine. Once we received an order we typically would deliver it 4-5 months later. When the time came and the machine was ready, I called to ask where we should ship the machine.

"When they bought that first machine from us, it marked their entry into tufting. I naturally thought we would be sending the machine to their plant in Winston-Salem, North Carolina, but the owner said, 'I don't know where yet – we're working on something else.' He explained that they were deciding on a site for the tufting plant. As it turned out, they wound up locating it in Ringgold, Georgia, rather than Winston-Salem," North said.

"There was no way we could warehouse a machine like that – we just didn't have the additional space – so the owner of Salem Carpet agreed to let us put the machine into storage for several months and he would pay the costs," North noted.

"One day we were having our monthly machine transported to Coronet in Dalton on a 'low-boy' tractor-trailer. I-75 had just been completed, and the 120-roll scroll, 5/32-loop pile machine was headed down the freeway. When the driver came off the Rocky Face exit, he probably was going a little too fast as he turned to go east on Route 41, toward the Coronet plant.

"The entire machine slid off the trailer, onto its side, hitting the pillars of the I-75 bridge going over Route 41. The impact smashed the machine into hundreds of pieces – it was a total loss. The bridge survived, but there was no hope for the machine – no way of repairing or salvaging it.

"Not long afterward, the phone rang in my office and it was Mr. Bud Seratean, president of Coronet. He had heard about the crash. He told me that no matter what, he expected to receive from us another machine the following week, just like the one that had been destroyed en route to his mill.

"I had no idea what to tell him. We had a full backlog of machines being built, each with a 4-5 month delivery time frame. Bud let me know in no uncertain terms that we had to do something. He needed to have a new machine the next week.

"Then it hit me – we had the same machine in a warehouse in Chattanooga, waiting for a delivery date and location from Salem Carpet Mills. I called the owner on the phone, asking if he would like to get out from the rent he was paying at the public warehouse. He immediately wanted to know what I was getting at. I proceeded to tell him the whole story about the wrecked machine and the sales order that had to be filled.

"It relieved me no end when he said, 'Let him (Coronet) have it.'

"I picked up the phone, called Bud Seratean and told him that he would definitely have another machine the following week. His reply was simple: 'I knew you could do it.'"

North recalled another transit-related incident that provided vivid proof that tufting machines and bridges do not coexist very well.

"In the early days, when we were shipping machines north, it required special routing for that type of machine because of its height and the less-than-desirable heights of certain old bridges along the major highways. That was before the Interstate system.

"We were shipping a machine to a carpet plant in Pennsylvania, and the driver had the proper routing. But along that route there was an old-fashioned bridge with a rounded opening. The center was the highest point of the bridge, and if the driver had gone down the center, he would have been okay. But he kept the truck to the right and the top of the bridge sheared off 100 of the yarn-feed rollers. They were a key part of the $200,000 attachment back then."

North pointed out the company did not incur financial losses at such times because the machines were fully insured, "but accidents still did not help in maintaining good will with our customers."

Northbound – for Denmark

If transporting these huge, sophisticated machines within the same region or even across the country presented challenges, imagine the difficulty of ensuring the uneventful delivery of the equipment to international destinations. Usually the machines still arrive safely, but North told about a memorable freight nightmare.

"Once we were shipping a machine to Denmark. Because of the height of the machine, fitted with pattern attachments on the front and rear, we had to put it into a crate, with the crate to have been lowered into a shipping container.

"The machine was loaded into the crate, but instead of being stored in the hold of the ship, by mistake it was left on deck for the entirety of the ocean voyage. The crate was not watertight, so by the time the ship arrived in Denmark, the machine had become fully awash with seawater. The salt and corrosion had rendered it useless to the company it was being shipped to."

Even though his company's obligation for the care and transport of the equipment officially ended at its loading dock, with the shipping agency assuming the responsibility from that point, North said such uncommon incidents still were frustrating. They hindered his company from providing the products and accompanying service in accordance with the promised schedule – wherever its destination might be.

Which provides a convenient segue to another important topic: How tufting became a global enterprise in the first place, exporting the wonders of carpet to homes, businesses and public places on virtually every continent. We will take an inside look at that intriguing development in the next chapter.

8

TUFTING GOES GLOBAL

The post-World War II fascination with tufted products, especially carpet, was not an isolated American phenomenon. As consumer interest in affordable tufted carpet was showing steady growth in the United States, initial measures also were being taken to introduce the wonders of tufting to other parts of the world.

Great Britain, where woven carpet had been a fixture among the elite for centuries, was at the forefront of this trend.

As Randall L. Patton writes in *Shaw Industries: A History,* "The British industry had followed a pattern similar to that of the American carpet trade in the post-World War II years. ... As in the United States, British demand for carpets expanded in part due to 'forces external to the carpet industry – to a low level of unemployment, rising real incomes, and a substantial volume of residential building.' A major contributor to this British boom was…the introduction of new technology, especially tufted carpeting."

British manufacturers were quick to embrace American tufting methodology and technology, and companies in England began producing sizable quantities of carpet in the early to mid-1950s. Virtually reversing the expeditions of the 17th century colonists, Cobble Brothers was eager to make inroads into this potentially lucrative overseas trade before the industry became too entrenched.

Becoming a Trans-Atlantic Competitor

Lewis Card recalled how this came about: "A man had started a business to build tufting machines in England. It was called British Tufting Machine Company – BTM. So we decided to start a competing business there.

"Jim Muse, a brother of George, was also a part owner of Cobble Brothers.

He came to me one day, not long after the end of the war, and said, 'Lewis, I want you to go to England and take a look around.' Jim added that he would like to go with me.

"We flew by plane to Blackburn, England, going by way of Manchester, which is where we landed. Once we got to Blackburn, we looked around and located a building where we could start a business. Then we went to London to research the region a bit more.

"It rained the whole time we were there – it was in the winter. The hotels were cold, the weather was cold and miserable, but I still told Jim that I would come back in a week or so and get started on the business. He said, 'That's fine, Lewis, but I'm not coming back.' He was from Sugar Valley, Georgia, and that was where he wanted to be.

"We followed through on starting the business in 1956, which meant I had to travel. Our primary competitor was BTM, and we eventually bought that company. Today Cobble is still in operation and owned by Spencer Wright Industries."

Opening the offices in Blackburn, England proved to be a success, serving as a gateway for Cobble Brothers not only to British trade, but also to the European tufting industry.

The Myth of International Travel Glamour

"I didn't find traveling internationally exotic. It was just part of the work," Lewis admitted. "If you have ever been to Blackburn, England, you wouldn't think of it as exotic. I learned to like the place, and I still do. But it's just an industrial town.

"What turned out to be an adventure was watching the business develop, which I really enjoyed. The British company had started building tufting machines over there, and I went over in response to that, making sure they did not gain a competitive advantage. We not only competed, we excelled, and were able to buy the British company six or seven years later."

The matter of getting to England in those days was a formidable undertaking in itself. Flying was difficult, long and tedious, hardly an exciting prospect, according to Lewis. "It was tiring. Flying from Chattanooga to Blackburn was a 24-hour ordeal. We would fly to New York City, then to Gander, Canada (the closest point to England from North America), to Shannon Island and on to Manchester, England and finally, Blackburn. Someone would pick me up and it was another 35-40 minute drive from there."

But his vision for expanding Cobble Brothers' international impact was not restricted to the United Kingdom. "I would not only go to the factory, but also to the Continent – to Germany, France and Italy – seeking to attract more business," Lewis pointed out. "All the days I was over there were 18-hour days. I would try to sell in the daytime, and then customers expected to be entertained at night. We didn't have to do that here in the United States, but it was common overseas. You had to entertain there to win and retain your customers."

When Lewis started taking trans-Atlantic jaunts, jet engines for commercial aircraft were unheard of. All the planes he flew on were propeller-driven, which sometimes injected a bit of drama and suspense to the travel experience.

"I made several trips to England, and in those days it was not uncommon for the airplane to lose an engine during the trip. It caused me considerable anxiety at first, but after the first one, it became routine."

Eating and sleeping in transit also presented challenges, Lewis said. "The food? It was all right – for airplane food. Hardly worth making a special effort to get!

"Sleeping was another matter. On one trip from Manchester to New York, I was assigned a berth for sleeping – and got it by accident. Someone had cancelled their flight, and the ticket agent told me she had one berth available and asked about whether I wanted it. I said, 'Yes, Ma'am!' It was the only time that happened for me."

Building a Business on 'Mr. Tufting'

A key component to Lewis's company's success in the European market was an enterprising gentleman named Edgar Pickering, who was on the scene about the time that Cobble Brothers, Ltd. was started in Blackburn. "We had sent one of our engineers to manage the company, but about two years later we hired Pickering to manage it. He was a good salesman and did a wonderful job of promoting tufting machines. He became known simply as 'Mr. Tufting' in the British trade papers.

"His role there was similar to mine in the United States, since I also was a salesman for Cobble, along with managing the business.

"Pickering had grown up in the textile industry, in loom manufacturing, so he had a strong background and a lot of experience. We first had him making parts for us as a subcontractor before he came to Cobble to manage the plant in England. He did a good job, both in selling and managing the company.

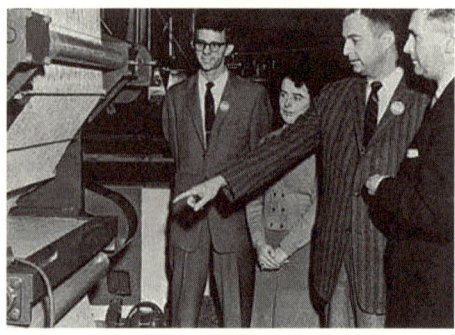

Observing the operation of a loop/cut tufting machine at an
international trade show in Manchester, England in the late
1950s are (from left) Roy Card, an unidentified woman employee
of Cobble U.K., Lewis Card, Sr., and Edgar Pickering.

"He was a strong individual. He had great faith in our products and
passed that enthusiasm on to the customer. He was a very aggressive salesman.
We gave him almost full control of the expansion of the Cobble business
into England and throughout much of Europe, and he proved to be a very
significant part of our growth."

Sharing the Global Exposure

After becoming established in responsible leadership roles with Cobble
Brothers, Roy Card began taking his turn in representing the company at
international trade shows. One year he spent six weeks in Lima, Peru for a
trade show and follow-up interaction with potential clients. That particular
year he also would spend 4-5 weeks in England.

"After Singer bought Cobble Brothers out, we worked with offices in South
America selling sewing machines," Roy said. "We tried to educate (Singer
leaders) on what a tufting machine was, but it seemed like the hierarchy at
Singer thought we could sell tufting machines about the same way as sewing
machines.

"I think the mistake Lewis made was in telling them the first tufting
machines were converted Singer sewing machines. So they apparently
presumed tufting machines would be easy to sell. Since all of the Singer
offices sold sewing machines, we would call on the same customers to try and
sell tufting machines."

It seems obvious in hindsight, but it took time to convince Singer's
leadership that the markets for sewing machines and tufting machines were as
different as the prospective buyers for MP3 players and home theater systems
are today.

Roy's experience with international travel taught him the wisdom of expecting the unexpected. "I enjoyed seeing different countries, but it gets tiresome after a while. In Lima, Peru in the early 1960s, they were having a mini-revolution, and I woke up one morning to hear gunfire outside my hotel. I looked out the window and saw soldiers running.

"I was told it was about teachers revolting over their pay. I thought it was really something, a matter to be very concerned about. But the next day when I mentioned it to a customer, he said it happened so often, it wasn't a very big deal for them."

Pursuing Expansion into Asia

Harold North's first international sales trip was January 1960, when he traveled to Japan representing Cobble Brothers. "We were being inundated with inquiries out of Japan. They were very interested in learning about tufting, and we felt the Japanese market was ready to break loose.

"Lewis popped into my office one day and said, 'I need you to get on the next flight to Japan.' Since then, I have been everywhere in the world that they had carpet industry or were contemplating putting it in."

North's sales excursions even took him behind the Iron Curtain, long before the demise of Communist rule.

Lewis Card, Sr. and Harold North (center) present a new
tufting machine to Steven Utermark, president of Uter Pty.
Ltd, a South African carpet mill, in the 1980s.

"In 1959, Cobble received a visit from a Russian delegation and we ended up selling them the first tufted carpet installation for that country. We sold them six machines and finishing equipment. We shipped the machines to them and were paid promptly.

"My first visit to Russia was in 1964. We had not heard anything from the Russians about installing the equipment – which we much prefer to do, to make sure it's operating properly. But their attitude had been, 'If Americans could do it, we can do it,' so they never accepted our offer to help.

"So on that visit there in '64, I asked to see some samples that had been made by our machines. No one seemed to even know where the machines were. Then, during another trip to Russia that I made in 1967, I was shown a two-inch square sample from one of the machines. It was the lowest of low quality. But that was then. Today, they are doing very good work there. We have shipped two or three new machines to Russia in the past few years."

'Sewing Off' and Trouble-Shooting

In his technical role with the Card companies, Jerry Hendricks was asked to travel to make certain the machines were installed and operating properly. He went out on the road for the first time around 1965 and spent much of the next 10 years "sewing off" machines, installing various attachments, and trouble-shooting for machines needing repair. His travels took him across the United States ("from Massachusetts to California – we did a lot of upholstery fabric machines back then"), as well as to Latin America, New Zealand, Japan, Canada and Australia.

The mechanical aspect of his work was predictable for the most part, but that was hardly the case with the different cultures he encountered.

"We had a situation when I was at a carpet company in Peru in the 1970s or early 1980s, Hendricks recalled. "I was working with Singer-Cobble, selling a controlled needle bedspread machine that cost about $450,000 at the time. My boss had sent me to Peru to meet with a guy who was buying machines in the textile business, sheeting and fabric. We had sent him a quotation, and it was finally time to do business.

"I flew to Lima, Peru and stayed at the Sheraton Hotel. We were working through Carlos Manacho, the Singer industrial sewing machine agent there. He was a native Peruvian Indian. The owner of the company was Roberto Ardillas, and Carlos drove me to his office to do the negotiations.

"When we got there, the agent took the pens out of his shirt pocket and put them into his pants pocket. I asked him what he was doing, and Carlos replied, 'Peligroso! Peligroso!' which is Spanish for 'danger.'

"Once we got into the owner's office, his daughter, Juana, who spoke

English, did the translating for us. After talking for a while, we agreed on a down payment. They then took me to a little office in the middle of the shop. I was sitting there and they brought in a brown bag. I opened it and inside were American dollars that I later learned had been obtained through the black market. I counted out $30,000 in currency, which was our agreed-upon down payment for the machine. They expected me to take this down payment, in cash, back to the United States with me.

"Keep in mind that I was young and naïve at this point. I thought I could just get the money and take it home. Then, as I started out of the shop, I realized I was being closely watched. Suddenly Carlos's words, 'Peligroso! Peligroso!' came to my mind. I got into the car and we headed back to the hotel. Finally it sunk into my head that I had $30,000 in cash, kind of hard to hide. I thought about putting the money into a hotel safety deposit box, but was becoming paranoid. 'What if the hotel employees are dishonest?'

"I didn't know whether to store the money in my hotel room, sleep on it, or what. I got Carlos to get me two money belts and I jammed the money into them and wore them around my waist. I thought about somehow having to protect the money for two more days in Lima, but then it dawned on me where my next destination was – Bogota, Colombia, the drug capital of the world!

"Once I got to Bogota, I had to spend several days there before leaving for home. Needless to say, I aged quite a few years in less than two weeks. I attribute much of my gray hair to that particular trip.

"Even when I got back to the United States, arriving in Miami, my ordeal wasn't quite over. On the declaration card you have to fill out when you come from another country, I had to lie because you weren't supposed to bring in more than $10,000. When I got back to my office at Singer-Cobble, I told my boss never again – I would not do something like that ever again."

Installations and Automatic Weapons

"Another time I was in Guatemala to help a guy named Roberto install a machine at his company," Hendricks recounted. "I was waiting for him to pick me up at the airport, but somehow our signals got crossed. Finally, I called him and discovered he was at his office. He said he had come to the airport, but didn't see anyone looking to be picked up. So I told him where I would be waiting and what I was wearing, and he assured me he would return to the airport as soon as possible.

"After a while I saw a little Chevy El Camino drive up, and in the back was a guy wearing military fatigues and holding an automatic weapon.

Roberto and a little boy, his son, got out of the truck, greeted me and we all got back in.

"They took me to my hotel in Guatemala City and said they would come back to pick me up for dinner at 7 o'clock. When they arrived, again in the El Camino, the military guy was with him again. The next morning, they picked me up at 8 to take me to breakfast at the family home. They weren't in the El Camino this time, but the little boy and the military man with the automatic weapon were sitting in the back seat of the car. Another peculiar thing was that the driver had a rear view mirror on a pedestal in the front seat that he could rotate as he wanted.

"At Roberto's home there was another military guy at the gate, also with an automatic weapon. Roberto's son ran up to the soldier and jumped into his arms. 'How strange,' I thought.

"Later, at his business, I finally got curious enough to ask Roberto what the armed military men were all about. He explained that his family owned the only brewery in Guatemala, and the year before his brother had been kidnapped for ransom. When he tried to run away, he had been shot and killed.

"Our next stop was to see the grandfather, apparently the brains behind the family businesses. There was a high wall all around the property, and a narrow driveway with walls on both sides. The driveway opened into a cul-de-sac where I saw two pillboxes with automatic weapons poised for use. Beyond that was a wide-open residential compound – it was like a walled city.

"The grandfather was a delightful man, obviously pleased to have someone to talk with. He was thrilled to take me on a tour of the brewery. But all the while I was thinking what a horrific thing it must have been to have to remove themselves from the real world because of their wealth and the dangers it raised. In that part of the world, at least, they were essentially captives to their own affluence."

The Reluctant Traveler

The late James Jackson, whose skills as a machinist in manufacturing parts for tufting were so critical to the success of Lewis and Roy Card's companies, reluctantly had to dabble in international travel for them – but only on one occasion. It was not an experience he cared to duplicate, according to his son, Sonny.

"In 1965, while Daddy was with Card & Co., he arranged to purchase one of the first lathes controlled by a punch tape. He went to Switzerland, saw the machine demonstrated, ordered it and had it sent to Chattanooga. Of course, to get to Switzerland, he had to fly in a commercial jet – which he hated. In the Air Force he had trained to be a paratrooper, but on one of his graduating jumps his chute did not open properly. The fall almost broke his back.

"For a long time, I thought that was why Daddy hated to fly, but then I found out it was because he didn't trust jet engines. The old propeller planes had been built to fly on one engine, even if they lost the other engines, and Daddy didn't feel jets had the same safety margin compared to the older planes."

Fortunately for the elder Jackson, his value to Lewis, Roy and their succeeding companies was in his machine shop, not on foreign soil.

Minimal Foreign Competition

In the 1970's and '80s, Japanese manufacturers began making major inroads into the American market, particularly with automobiles and electronic equipment. Even with this great industrial growth, however, Japan did not mount any significant competition for carpet tufting machine manufacturers. And that was by intent.

"They were looking to make high-volume machines – for sewing, knitting, automobiles. They were not interested in making machines that might sell 100 units per year at most," North pointed out.

"We were – and continue to be – specialists. We know how many machines we can sell, and all the technology is here. For someone else to start up a major carpet tufting machine company – having to develop the technology and acquire the necessary expertise to produce machines in very limited volume – does not make much sense economically."

So, similar to their business in the United States, companies led by Lewis and Roy Card enjoyed a strong competitive advantage outside the U.S. borders. The fruit of their hard work, initiative and innovation remained in high demand in many parts of the world.

Variety without End

"An almost endless novelty" was how one high-ranking carpet executive described the product of tufting machines by the 1960s and early 1970s. Development of new attachments, coupled with design modifications to the machines, enabled manufacturers to produce carpet in a seemingly infinite range of styles, colors and textures. This variety helped to maintain consumer interest, disproving observers who regarded carpet as a "passing fad," which is what tufted bedspreads and robes had proved to be.

But where did all of this creativity, this inventiveness come from? What inspired and drove men like Lewis and Roy Card, as well as their peers, to continue exploring and expanding the capabilities of their machines? Some of it was necessity – a simple matter of seeking solutions for problems that arose, or striving to satisfy the needs of customers who would ask, "Why can't you make a machine to do this?" Some of it was curiosity, a desire to experiment and see if a certain novel effect could be created. As in any creative, innovative process, some of it also resulted simply from people working together, bouncing ideas off one another until they came up with something fresh and new.

This is where Lewis and Roy were particularly adept, as we will see next.

9

CARD = CREATIVITY + AMBITION + RESOURCEFULNESS + DETERMINATION

Advances with the tufting machine entailed a complex, multifaceted process. It was not simply a matter of making the machines wider and adding pieces, because bigger also meant heavier and broader, putting the laws of physics to more stringent tests. Although the principles of tufting remained fairly simple and straight-forward, the how-to of applying them to the latest advancements became increasingly involved.

An example was development of the needles and other corresponding "gauge parts" – the reciprocating knives and loopers that functioned from the opposite, lower side of the backing material. Machine manufacturers adopted the term "gauge" to describe the spacing between the needles. More precise, more durable gauge parts had to be developed as the spacing became more narrow along the needle bars.

As Max Beasley noted in his writings, "This required a continuing development program, not only to reduce gauge but also to improve cutting efficiency... At (one) time the finest gauge was about 3/16-inch. Some of this fabric produced on 45-inch machines was joined together, sometimes with an adhesive strip to make a piece of carpet wider than could be produced on a machine. Lewis was wanting a 'broadloom' carpet machine for Cobble... we thought we were only talking about a difference in size. How difficult could that be?"

In actuality, it could be – and was – *very* difficult. Roy Card summarized it this way: "The tufting industry is very simple really – that is, simple when you have only one needle and one looper or hook. When it gets complex is when you have thousands of them all working together at the same time. That means you have to have every one of them almost perfectly on gauge for all of them to work together."

Lewis, from his perspective as the even more-seasoned brother, concurred: "Prior to World War II, I was working on single-needle machines. Today some of them have as many as 2,500 needles, but the basic principles of tufting were there from the beginning with the single needle. Since that time a lot of development has been added – but the fundamental principle of tufting was there in the single-needle machine. It's always continuously been the same: You have a needle, put it through cloth, grab it with a looper (or hook), cut it or loop it, and control the feed of yarn."

Uncharted Tufting Territory

Part of their challenge was the fact that they were true pioneers, exploring uncharted territory in the new world of tufting. As Roy pointed out, "When we started, there was no 'tufting machine industry.' We didn't have any background or previous history, so we couldn't go back and say, 'This is how we did it last year.' Because we didn't have past experience to draw from."

Lewis agreed: "We were fortunate – everything we did was a new invention. Nobody had ever done it before, and we had to do everything, coming up with new developments to solve problems and create new ways of doing things. It was exciting to even have the possibilities and opportunities to work in the industry.

"There was not some grand plan that one day the industry would arrive at where it is today. There was no business plan for how to go from bedspreads to carpet. But when we moved from bedspreads to scatter rugs, it became easier to see how the industry could go to carpet. Of course, even though it was easy to see – it wasn't all that easy to do!"

The "creative team" of Lewis and Roy never arrived at a point when the development process became simple or effortless. However, their unique synergy served as a testament to American ingenuity, determination, know-how – and cooperation. Perhaps having learned valuable lessons by observing the professional friction between their uncles, Albert and Joe Cobble, Lewis and Roy were able to avoid such clashes. And for that, the tufting industry owes a debt of gratitude.

It may have been that, unlike the Cobble brothers, the Cards succeeded in accepting their respective strengths, understanding how they could complement one another in ways that resulted in the whole becoming greater than the sum of the parts. Looking back on those times, they offered some viewpoints, as did some family members and individuals that worked closely with them.

Lewis observed, "Roy and I were not geniuses, just hard workers. We have a lot of patents, but we had to dig those (ideas) out. It was not like a brilliant light suddenly turned on."

Several people close to the brothers agreed they both were "great concentrators and had tremendous persistence" in seeking to solve problems and come up with new, necessary innovations. Asked about those descriptors, Lewis replied, "Does 'very persistent' mean hard-headed? I am. Roy is not too much that way. But when people would say, 'I'm not sure that will work,' we would respond, 'Do it anyhow. If it doesn't work, do it over.'"

As for being persons of "great concentration," he said, "It's a matter of not letting an opportunity get away from you. If you get an opportunity, grab it and stay with it. Persistence might be a good term.

"Larry Gable, one of the fellows we had at Tuftco and Card & Co., often said it was like we had blinders on. We would go for what we wanted and not let anything on either side distract us.

"Anything we developed, nine times out of ten, it was not just putting it out there and seeing it work. We would have to do a lot of revisions to make something work – but we were willing to do whatever was necessary to get it done."

While Lewis contended they had similar personalities, Roy acknowledged they were different enough to provide necessary balance. "Lewis was a little more forward-thinking. He terms it 'aggressive.' I was a little more conservative about spending money, and putting up buildings. He had more confidence than I had."

"I never thought Roy had a problem with confidence in what we were able to do – or could do," Lewis countered. "We all were thinking about the future, but yes, I'd say I was more aggressive. Roy was probably more mechanical than I was. He was really good at getting a machine into operation."

"I was more concerned with the day-to-day operations," Roy stated, "and Lewis was more geared toward the future. I was always worried about how to get that machine running properly, and about developing machines for accomplishing different purposes."

"I was more into developing machines for different fabrics, but actually, we both did some of both," Lewis recalled.

Uncovering How Things Worked

The quest to comprehend the how's and why's of things served as a driving force for Lewis. His older son, Lamar, established a very different career as a commercial film producer in California, although he did work in the Cobble Brothers shop as an adolescent. He retained vivid memories of watching his father mentally dissect some object as he worked toward a solution for a problem.

"Even before Roy came into the company, my dad had been laying the groundwork (for developing and building tufting machines) with Joe Cobble. Dad has a very unique ability – I call it understanding how to break a thought-set. He could look at something and see it as something else, or as being a part of something else."

In other words, Lewis could step outside of a paradigm – the standard, typical way of perceiving something – when he looked at it.

Lamar Card (left), older son of Lewis Card, Sr., with Edgar Pickering, who managed Singer-Cobble's European operations in the 1960s.

"He had the ability to see a wire, for example, and envision it as something else, perhaps to make it into a piece" of a device he was trying to develop at work, Lamar said. "He was always playing with spoons or salt shakers at the dinner table, trying to see them as something else. It drove my mother crazy."

Being engaged in a creative industry himself, Lamar viewed his father's gift as more than simple mechanical know-how. "Just like an artist that takes a common image – one that is readily recognized by many people – and shapes it into his or her personal form. Actually, I once called Dad an artist, but he didn't like that at all. But at the core, he is a very creative person, highly focused in a mechanical sense."

"I'm not certain how he passed that on to his staff, because I never saw him as a particularly good teacher. But it was amazing to see him and Roy work together solving problems."

Lamar's younger brother, Lewie, affirmed the observation about his father's curious, analytical mind. "He and Joe Cobble were innately curious about how things worked. They always had something in their hands, looking at it."

Lewis Card was a "quick study," according to his older son, adept at grasping lessons about life and work without having to relearn them. He not only adapted from personal injuries (losing sight in one eye as a boy, and a portion of a finger as a young man at work), but also was able to master mechanical principles and readily apply them to tasks at hand. "That (capability) always amazed me," Lamar said.

But even within his own family, Lewis did not have a corner on the market for curiosity and in-depth exploration. Roy remarked, "My wife has always laughed at me, saying anywhere I go, I get down on the floor and look at the carpet to see what kind it is and how it was made. It doesn't matter whether it's a motel, hotel, restaurant – I try to figure out for sure how the carpet was made, how the pattern or design in the carpet was produced."

Not 'Hindered' by Academic Training

After high school, neither Lewis nor Roy received any formal academic training in the work they did, but that might not have been a bad thing. As their sister, Bett, reflected, "I often wonder if they had gone to college, and perhaps had become mechanical engineers, if that would only have slowed them down in trying to figure out how things worked and could be done (in developing tufting machines)."

Roy responded that there might be some truth in her observation. "We never could hire professional engineers to do what we did. When we consulted with engineers, they would say that what we were doing wouldn't work – until we got it done and showed them it would work."

Charlie Monroe, Roy's son-in-law (married to his daughter, Renee), gave his perspective on the differences between the two brothers: "Lewis owns two houses, and each has a garage full of tools. He is very mechanical, and likes to look at all the possibilities for solving a problem.

"Roy is stronger in the application of technology; discovering the various applications of new innovations for tufting machines and the manufacture of carpet. Roy typifies the adage that 'true inventors never get through inventing.' He is always trying to perfect his inventions, trying to make them better – even today."

Larry Gable used an anecdote to draw a contrast between the brothers. "One time we were having major difficulty in getting some parts made. We had been working for hours on a Saturday when Lewis stopped by. He asked to see what we were doing, and then calmly made three suggestions for minor changes. Then he left, saying he had to be at a meeting in Atlanta. We used two of his suggestions and our problem was solved.

"He was kind of like the Lone Ranger in that way, strictly a behind-the-scenes guy."

Roy, by comparison, according to Gable, was particularly good at working with different kinds of yarns and backings to determine how best to serve the tufting industry. "He would say to Lewis, 'This is a problem we're having in the mill,' and Lewis would work to find a solution. Roy was good at interacting with the yarn and backing people in the mill, while Lewis would talk more about the machines themselves. They meshed together perfectly.

"Roy was more of a hands-on guy, even with me as a shop operations manager. We would communicate with Lewis more infrequently. But he had an engaging way about himself," Gable said.

"One time Roy and Lewis were planning for the next year and asked us to draw up a list of the equipment we would need over the next 12 months. I went to Roy with my list and he okayed it. As Roy and I were leaving his office, Lewis came out of his office and asked if Roy had approved everything on my list. I said that he had. Lewis looked at my list and chided Roy, 'You approved everything on his list? He didn't even want the first three things – he was just trying to see what you would say!' Lewis – you just loved working with that guy."

Valued Mentors in Their Craft

Allen Neely, who worked with Lewis and Roy for more than 25 years in various capacities, spoke of them from the perspective of a grateful protégé. "In my time with Lewis, I spent a ton of time on tufting and developing gauge parts – our goal was to make our machines user-friendly and easy to maintain. Just as he had provided an opportunity for Roy to grow many years before, he did that for me as well.

"Working with Roy in the sample department, we would find applications for new technologies we had developed. For me, he was not so much a boss as he was a mentor. I still keep an article I first saw 15-20 years ago entitled, 'Having a good mentor is critical to a successful career.' At the same time, Lewis was also a mentor to me in every respect. He taught me the ethics of work and he was patient enough to explain historical precedents, why things were successful or why they failed. I was very mechanical by nature, so between Lewis and Roy, I learned the art of tufting.

"Lewis taught me to weigh and make decisions. He gave me my first big raise, and invested in me personally. And he opened doors for me to progress through the business, enabling me to advance from a machinist to a management position with the company. I feel that even after he became inactive with the company, he was still promoting me. When you can have time like that with the chief and develop a relationship, that's rare."

Jerry Hendricks expressed similar feelings. Since Lewis's career was beginning to wind down by the time his own was getting into full swing, Hendricks said Roy was his greater influence. "I consider Roy my mentor. He gave me my opportunity in this business, helped me to learn about the machines – especially the technical aspect of it. He was very patient, always trying to instruct me. He has a tremendous analytical mind, and doesn't make rash, quick decisions. He gathers all the necessary details and likes to discuss the issue before any decision is made about how to deal with a problem or rectify it."

The Key – Finding the Right People

There is a time-honored saying that whether in life, work or sports, "you win with people." While Lewis and Roy were always setting the pace in their companies, they understood well the importance of surrounding themselves with talented people.

The late Harold North, who in 2007 celebrated 50 years in the tufting industry, primarily in sales roles, grew to appreciate their discernment when it came to personnel.

"Both Roy and Lewis were good at 'debriefing' their technical people. They learned quickly who their best people were, and eventually it got to the point where there were not many left in that department who weren't very good people technically speaking. If you didn't know what was going on with the tufting machine, and didn't demonstrate a capacity for learning it, you were not working for Roy very long. He was patient with the growth of his employees, but you had to show that you were growing.

"This wisdom and expertise came from their being out there, getting down and dirty, doing the job themselves and knowing what it took. They would get in and come out as greasy as a hog, if necessary. They were not the polished, got-to-have-their-nails-done style of managers – they would just dive in and do whatever it took to get a job done right."

Hendricks credited their ability and willingness to allow their employees to learn from experience, through trial and error, rather than micromanaging the work. "There were many times when Roy would let us young guys make mistakes, and go through the learning curve, when it did not affect customers.

"On mechanical jobs in the plant, when we were building a machine for a customer, for example, Roy would be specific and direct. But in the sample department, he was not so firm and hands-on. He would allow us to learn what we needed to know about the technical aspect of the job.

"In assembly, we knew that the machine going out was the customer's first impression with what was going on with Cobble Brothers Machine Company. It took on a different priority than learning how the machines worked in the sample and development area. We didn't want anything going out that would indicate that we were less than the best," Hendricks recalled.

"The sample and development department was important and scrutinized, but not with the same focus. We were there to train and learn on our own what was needed to fix a machine. Roy would leave us alone for a time, but after a while, if we were still working and trying to figure out a solution, he would come out and tell us what we needed to do – and why. Afterward we would wonder why we didn't think of the solution on our own. Roy always wanted you to dig for answers and grab them, rather than having them given to you.

"He always asked very pointed, deep questions. He wanted you to think about the questions he was asking and be able to answer with something that had meat on it, to show you were learning about what you were doing – not responding with just a 'yes' or a 'no.'

"At times he was kind of hard, but later you would learn to appreciate what he did and the way he did it. I have always had the utmost respect and admiration for his analytical mind. He's a special guy. You could go to him and ask a question, and he might say something like, 'I don't know the answer to that right now.' An hour later he would come back and suggest, 'What if we did it this way?'

"Roy liked to demonstrate that if you ask enough questions, you will get a better answer in return. He would teach us how to draw out information that we needed to solve a problem. It would be embarrassing to have him ask, 'Why didn't you ask this question (of the customer)?' He didn't do that to embarrass or humiliate you, but to teach you how to thoroughly think through the process.

"Of course, Roy got a lot of this from Lewis, who mentored him. Their approach to problem-solving was very similar in that respect. Roy was a firm believer that the more good input you had, the better your solution to the problem would be."

The Right Answer at the Right Time

Gable agreed with North and Hendricks about Roy and Lewis both being able to find the right person to do a job and then support that individual to the fullest, but said Lewis had one particularly notable quality that he greatly admired. "He had an uncanny ability to develop equipment to help the industry advance. He was very astute in anticipating the needs of the tufting industry."

Gable gave an example: "In 1965, Lewis approached me and asked what I thought about building a better, more lightweight mending machine. At the time, mending machines (used to repair flaws as carpet was being tufted) weighed 12-14 pounds and they were all handled by men. But women were starting to come into the mills to work, so a smaller, lighter mending gun was needed.

"I developed a little machine – it weighed four to four and a half pounds, and gave it to one of our salesmen to show to Lewis. He called me back immediately and asked, 'What do you have in the chamber?' I told him that there was nothing in the housing, but the air chamber would capture and redirect the exhaust air from the motor.

"'Why don't you try to make it a little smaller by reducing the size of the housing?' was all he said. That ticked me off, since I felt we had done such a good job, but it motivated me to do even better. The result was a two-pound, three-ounce mending gun which is essentially the same one that is used in mills today. Over the years, we have sold those things like popcorn."

Particularly memorable to those who worked with him was Lewis's versatility. One minute he could be involved in a technical interaction like the one described above, and the next he could be engaged with movers and shakers in the business world.

"The amazing thing about Lewis was that he could get right in the middle of things at the shop, then clean up, put on a suit and tie, and go out to talk with bankers and owners of carpet mills, being able to go toe to toe with them in complex business negotiations," North said.

"Lewis was a well-rounded businessman, very capable in many ways. He could be speaking with financiers in the morning, then spend the afternoon at the mills, interacting with everyone from owners all the way down to machine shop people and floor sweepers. He could relate well at any level, and talk to everyone with their same language. On the floor of our company, he would interact with managers, talk to workers about their jobs, and people who were building machines for him.

"He was very reserved, but he made a very good impression after people got to know him. Lewis was intense, focused and methodical – very devoted to his work and to the industry in developing and manufacturing tufting machinery. He was very serious-minded, not much of a joker."

Solving Problems, Responding to Needs

"Where do you get all of your ideas?" This question is often posed to creative people. For the enterprising Cards, the answer was simple. They were either working to solve problems or responding to the needs of their customers. As Bill Wiegard, an innovative executive in the floor covering division of Collins & Aikman in Dalton, observed about the industry in the 1960s, "little inventions came about because of necessity."

"We forever had people bringing woven materials into the shop, telling us they wanted to do that on the tufting machine, only 10 times faster," Roy said. "And a great majority of those things we eventually were able to do."

Lewis added, "A lot of our patents were simply solutions to resolve problems. But when there wasn't a problem to solve, Roy and I had a lot of discussions about possible innovations.

"For instance, hobnail bedspreads (which featured three-stitch tufts spaced 1¼ inches apart in both directions) were popular, and made by hand. But we figured that if somehow we could make the machine jump stitches, we would have a hobnail bedspread machine. So we worked at it, and one day we had what amounted to a Catherine Evans bedspread machine – we did it to develop new business."

Enhancements have been two-thirds of design and growth, according to Neely. This is part of the Card philosophy that the "customer in the field is the No. 1 priority" philosophy they have always maintained. Seeking solutions to problems in the field also prompted design changes. "We would always ask the customer, 'What's not working for you? How can we improve the longevity and reliability of our machines?' Invention is truly building on previous technology," he said.

This stimulation of searching for something new, or better, not only spurred innovation but also helped to maintain the Card brothers' enthusiasm for their work, according to Lewis. "A customer would come in wanting something done in a hurry, and that is what all your research is for – providing something new to sell to the customer. The best part of being in that business was the opportunity to develop it."

The Smoke (and Idea)-Filled Office

Neely said he will never forget the image of the shop's creative nerve center in the early days. There was a small, fabricated room in the middle of the shop that Lewis, Roy and he used as a working office. It had mirrored windows so that no one could see in, but contrary to the prevailing opinion, the one-way windows were not for spying on the work going on in the shop. Lewis just wanted to be able to work without distraction, while being able to look into the shop when necessary.

"This office had shelves around all four sides that served as a trophy case for little models of small tufting parts – trinkets, pieces of technology, samples that we had been working on. We never threw anything away, even things that didn't work.

"I smoked at the time, and the daily memory I will always carry with me is of Lewis with a chaw of tobacco in his cheek, a True Blue brand cigarette burning in an ash tray and an A&C Grenadier green cigar, and Roy constantly smoking a pipe. The office would become so filled with smoke that you couldn't see across the room, and all the while we were in intense discussions about tufting.

"Roy had a unique way of smoking a pipe – it was a continual puff, puff, puff as long as the pipe was in his mouth. He did this while thinking through a machine problem that needed to be solved."

The tradition of the "smoke-filled office," however, came to an abrupt end, according to Neely.

"I remember about 5 a.m. one morning Lewis called me at home, when I was 28 or 29 years old. He had been taken to a local hospital with angina pains. He said, 'I won't be back. I want you to take what we have been working on and show it to Roy this morning.' He later had to have heart surgery. Instantly the responsibility for research and development fell into my lap, and I ran the department for several years.

"When Lewis stepped aside, I quit smoking in his honor, in respect for the surgery he had undergone and his subsequent recovery. I felt that close to him. Roy did the same, and instantly that office in the middle of the shop went from a smoke-filled room to nothing, no smoke at all from any source."

Neely also described Roy as a "Scotch tape person" in his thinking process, particularly when pondering some perplexing matter. "When he came into your office, he would take a piece of tape from the dispenser on your desk and roll it around in his fingers while he was thinking. Also, when talking with someone, he would place a foot on a toolbox or desk and jingle change in his pocket, again just a little habit he had while he was thinking."

Max Beasley wrote that more than once he marveled at how both Joe Cobble and Lewis utilized their unusual gift of visualization to advance the innovative process. "I can't imagine visualizing a whole machine just in your mind."

Beasley compared this process to drawing, or poetry: "I have to begin drawing, and then the plan sort of unfolds as I go. ... I reckon Joe Cobble visualized the whole (yardage) machine in his mind, and he and foreman Sid Manning built it with Joe instructing Sid what to do. I never saw Joe work with his hands. … Lewis was good at visualizing too and he had a brilliant mind for designing, but was not able to put it on paper…"

Lewis talked about how he and Beasley frequently collaborated to bring an idea into reality. "Whenever I had envisioned a new part or attachment consisting of a group of parts to be put on a tufting machine, I would have Max draw the part to be manufactured. We would either do that, or find a machinist who could create the part based on how we described it."

More often than not, that machinist was James (Jack) Jackson, who played a prominent role in the Cards' later companies, particularly CMC.

"James was a machinist who worked for us for years," Lewis said. "If I wanted to make a machine part, I would go to James with a sketch – maybe on the back of a napkin – and tell him I wanted him to make this part. That was all he needed to know. I would give my sketch to him, and then go about my business. It's hard to find people like that – now or at any time. He was a unique individual."

James Jackson, Machinist Par Excellence

Even the best ideas are of little value if they cannot be executed. This was why Jackson proved to be such an indispensable asset at most of the companies the Cards owned and operated. He first worked for Lewis as the machine shop supervisor at Cobble Brothers, then moved to Roy T. Card & Co. to join Lewis and Roy at their Riverside Drive location, heading a crew doing metal machining. Later he and his son, Sonny, started their own machine shop in LaFayette, Georgia, making gauge parts and pattern slats, a mechanical means for creating carpet patterns with tufting machines.

"Jack was one of the best hires we ever made; he was an expert in tooling and machine fixtures," Lewis said. "He was a specialist in developing shop equipment to make specific parts, once prototypes had been produced. He was the best we ever dealt with.

"The Jacksons were very innovative people. To do this kind of work required experience, being a top-notch machinist, and being very innovative. It's different from being an average machinist, because you're always looking for new, better ways of getting something done.

"Sonny and his son, Heath, possessed a lot of the same talent, carrying on a family tradition of sorts. As with his dad, I could go to Sonny with a need for a specific job, or a specific part, and he would come up with the best way to manufacture that part."

James "Jack" Jackson was instrumental in designing and producing precision gauge parts for tufting machines.

Roy added, "They were doing the same thing with machine parts that we were doing with tufting machines; trying to find ways for correcting problems."

This talent became increasingly important as speeds of tufting machines increased, along with the demand for closer tolerances, as Hendricks noted.

"Then we were using 5/64-inch gauge for loop and cut pile, meaning there were 12.8 needles per inch in a gauge machine. That was a pretty fine gauge for that time. If we could keep the tolerances of our machines within .050 inches, we thought we were doing good," he said.

"Many of the adjustments were made by mechanics that literally had to bend hooks by hand for the machines to run. However, as machines tightened in gauge, the need for accuracy became more critical. We had to get better in making gauge parts to eliminate all problems. Inaccurate gauge parts contributed to rapid wearing of needles, knives and hooks, and it was costly and time-consuming to have to stop a machine and change out parts.

"Today, we deal within .005 total accumulative error tolerance from one end of the needle bar to the other. This has greatly increased needle, looper/hook, and knife-cutting life. We have succeeded in improving efficiencies of tufting machines across the board, regardless of gauge or construction."

'Mechanical Sherlock Holmes'

The elder Jackson truly set the bar for producing quality machine parts and did not take that role lightly.

Sonny Jackson recalled that as a boy, he would often see his father preoccupied with doing mechanical detective work, even during leisure time. "He would read a machinist's handbook rather than reading a book or magazine. If there was something he encountered during the day, whether it was not knowing how to make a part or how to fix a problem, he would look for the answer in the handbook, study it and be able to solve the problem (when he went to work) the next day."

It often required mathematical know-how – an equation, logarithms, or some other numerical calculations – to solve the problems, such as finding correct gear ratios, he said, "rather than having computers do it for you, as they do today. In order for gears to work together, they must be perfectly timed; they have to be in perfect synchronization for tufting machines to work with the precision required.

"But in the early days, there was no going to the book or an instruction manual. You just had to figure out equations on your own, taking general mathematical principles, laws and formulas and applying them to the design and manufacturing of machine parts. My dad was a master at that."

Interestingly, "Jack" Jackson never got very involved in how a completely assembled tufting machine ran. His interest and focus centered on designing and making the best possible machine parts so the machines using them could run at optimum capacity and efficiency.

Jack's ingenuity grew out of his intense desire to excel, his son pointed out. "He was a perfectionist, determined that whatever he did was done as well as he possibly could. I remember a time when he was at either Cobble or Card & Co., and his supervisor asked him to create big carts to roll carpet away from tufting machines. Daddy had earned the reputation that if there was a problem, 'Jack' would be able to fix it. But before Daddy could even get started, someone else quickly built a cart that had bicycle wheels on it. The supervisor asked, 'Why didn't you come up with that?'

"Then, with the carpet loaded on top, they tried to take it around a corner and the cart wheels locked up. Daddy just looked at the cart and said, 'That's why I didn't do it that way.' Then he suggested a practical way to solve the problem of the wheels locking up."

Despite his intensity on the job, however, Jack occasionally exhibited a playful side, according to his son. "He did put up with a little bit of horseplay at the shop. He had a good sense of humor.

"I used to hear a story about one of the men in the shop who bought a

Volkswagen, drove it to the shop and started bragging about what good gas mileage he could get. So as a prank, Daddy and another guy started putting extra gas in the car while its owner was at work in the shop.

"This fellow was amazed at how he kept getting better and better gas mileage in his car. One day he even commented, 'I bet it's been three weeks since I had to fill up that car!'" (Of course, that was in the days when gasoline prices were calculated in cents rather than dollars.)

If at First You Don't Succeed...

When we think of inventors, Thomas Alva Edison often serves as a classic example. In his endeavors to realize his dream of creating the electric light bulb, Edison is said to have commented that each failure simply revealed another way that his idea would not work and brought him closer to the ultimate solution. While Lewis and Roy may never be accorded such prominence in inventive lore, they too grasped the truth that the shortest path to success often passed through failure.

"We had a lot more failures than situations where things worked well initially," Roy admitted. "We failed many more times than we succeeded. A lot of people are afraid of failing, but you can't have a fear of failing if you want to succeed at doing something worthwhile."

Lewis agreed: "We never let anything that didn't turn out as we thought it would prevent us from doing something else. That was a big part of our business – developing something new and better. Being in a new industry there was always plenty to work on – everybody was in the same boat.

"For instance, we developed a nine-foot machine for bedspreads, but initially we had a lot of problems with it and had to go through a lot of changes and improvements to make it right. Nothing was ever easy. There just was no precedent, nothing to fall back on – we were out front, leading the pack.

"We always felt that if we had an idea and worked on it, eventually we would produce something from it. The good thing was that we did not have anyone to answer to, so we didn't have to worry about failing."

"I think we were positive, but realistic, too," Roy recalled. "We didn't try to paint a rosy picture on everything. It's just that we didn't think it was the end of the world if things didn't turn out as we thought they would. We'd just 'drop it and keep moving,' as Lewis liked to say. An example was tufted wall covering – that idea eventually didn't work out, but we were learning as we went along. We would either keep working at a problem until we could fix it, or go some other way (that proved to be better)."

One of their "secrets" was not demanding a high success rate from themselves. Instead, they wholeheartedly ascribed to the "if at first you don't succeed, then try, try again" philosophy. "We basically felt that if you got two ideas that worked out of ten, you were batting a pretty good average," Lewis pointed out.

Hendricks, who on the scene witnessed many of these failures – as well as the notable successes – confirmed this attitude. "They always had that quality of 'never give up, never say no.' If we concluded that something could not be done the way we were doing it, we would just try to do it a different way. They always had a very open mind to different approaches for problem-solving."

The determination to snatch success out of the jaws of failure energized the Cards. "I have enjoyed it from the first time I got the thrill of correcting a problem," Roy remarked. "I remember that particular situation was when we were having a problem in working with a new machine and were able to correct the problem using an approach never tried before.

"I don't know if I can describe the thinking process involved. You just think about the problem, try to arrive at a possible solution, and then start sketching (the mechanical solution). Over the years I have filled dozens of sketchbooks – I still have them. The fact is, 99 times out of 100, what you try amounts to nothing. But it's that 100th time that you are always shooting for.

"I have always had a desk filled with different gadgets and parts. Lewis is the same – we just like tinkering."

Roy recalled many times when customers would bring a sample of woven carpet and ask, "How close can you get to making this (by tufting)?" "Often we would explain we were not able to do much with it right then, but would see what we could do. And then we would work and work, trying to get it done.

"Over the years we found that solutions for many of our most effective patented items were simpler than we had initially realized – we just hadn't thought of it earlier. We would just keep trying, maybe doing something different, until we solved the problem. Sometimes it involved making changes to an existing machine; sometimes it required a new attachment, or taking a new approach."

That was indeed the case, according to Lewis. "Many times solutions we tried were more complicated than they needed to be. In designing equipment, for example, you're inclined to make it heavy because you don't want it to break. But if you are trying to get speed out of a machine, you work at reducing the weight of the moving parts.

"It's not the engine that causes the noise, but the vibration of moving parts. In a tufting machine, with needles going up and down, and loopers and knives going back and forth – and the bars that carry them – there's potential for a lot of noise. So you refine them, seeing how you can make them not as heavy as you originally thought they needed to be."

Patents: Not Necessarily Rocket Science

Through the years, Lewis and Roy were rewarded with more than 100 patents each, U.S. and foreign, for their curiosity and inventiveness. But as they have already pointed out, discovery rarely involved some kind of mystical process.

"One thing that surprised me, talking about patents," Lewis said, "is that often you think of something really basic, but something that no one has ever done before. I would say that 90-95 percent of new patents are a combination of old ideas to perform new tasks. Basically, the patents resulted from finding solutions to a problem."

For more than 50 years, brothers Lewis (left) and Roy Card have
been partners in tufting innovation and development.

Roy received his earliest patent in 1961; the latest was in 2006. "A lot of people ask me about getting patents, and I tell them what I know – that the only good way is through a patent attorney, and it's a long, complex process," he said.

"Only about three percent of all issued patents ever make any money, and only 10 percent of them ever get into the marketplace. With our patents, by working with a machine and the ability to run carpet (to test our inventions), we could see the fabric and the finished product even before we made the initial application for a patent. So, unlike many patented inventions, we had already seen the result and knew that it would work."

Who's the Boss?

While the Cards had their share of disagreements, as any two people would have when working closely together day after day, they never contested over authority. Roy asserted that Lewis "was always the boss," although the older brother retorted, "The only reason (I was the boss) was that I got there first." Then he added, "That was earlier. Now it's the other way around. Since I retired, he's the boss."

Hendricks pointed to their ability to share leadership responsibilities according to their respective strengths. "From the time I started working with the Card brothers, Lewis was basically in the front office, while Roy was in charge of technical service. But of course, a lot of what Roy taught us about machines came from what he had learned from Lewis. They both were extremely knowledgeable about the tufting process, very hands-on guys."

A key, it appears, was their ability to maintain a humble perspective. Despite being an integral part of creating a foundation for what has become a truly amazing industry, Lewis regarded his accomplishments with a shrug. "I was the luckiest man anywhere in the world. I was able to get in on the ground floor with a brand new business – not only in the United States but also the world. Developments came along as you could see new possibilities and applied those to what already was being done."

"Seeing the industry grow – and being a part of it – was one of the most satisfying things that happened to me, from a business and work standpoint," Roy added. "So many people were trying to get machines to do the things we thought they could, and the inventive part of figuring out how to get these things done was very satisfying."

Working with both of them, Hendricks shared in the exhilaration of being part of an emerging industry where everything seemed fresh and revolutionary. "There were a lot of things being developed and tried in that time and particular era. It was very exciting for me. There were always new things coming along, changing the tufting machines and what they could do. This involved the raw materials we used, as well as the machine itself.

"From its infancy to a more mature state, being there with Roy and Lewis at that particular time was very beneficial for all of us that came through it. It's like building a house – you have what you think is a great foundation, only to find out it could be better yet. In the same way, we were learning how strong the machine frame needed to be – heavier, with beefier frames, sturdier and with more precise gauge parts as we progressed from bedspreads to rugs to carpet."

The Card brothers were quick to recognize that the advancements in the tufting industry ultimately were the result of the hard work of dozens, even hundreds, of talented and dedicated individuals. Although Lewis and Roy were fortunate to play key leadership roles for decades, many people deserve to share in the credit for accomplishments that helped to develop tufting into what it is today.

"We certainly were not the only ones working to expand the capabilities of the tufting machine," Roy acknowledged. "There were multitudes involved in the development of the machine. We could cite people like Wallace Hammel, Rodney Hill and James Jackson, but there are so many that should be mentioned, trying to put together a complete list would probably result only in leaving out more people than we mentioned. Once you start naming names, you've got problems because of all those that you would unintentionally omit."

As a new decade – the 1980s – arrived, a new chapter for the carpet tufting industry was about to open. This would involve a changing of the guard and the introduction of some new individuals, representing the next generation of the industry. This would result in further solidifying the remarkable legacy of the Cobbles and the Cards.

Russell and Mae Card descendants, gathered for a 2007 reunion.

10 CMC – A New Company is Born

For more than 20 years, the carpet industry had enjoyed remarkable, continuous growth. The esteemed "experts" who had dismissed tufting as a "fad," predicting the market soon would return to the traditional woven fabric as consumer enthusiasm subsided, had long since conceded their errors in judgment. Nevertheless the adage, "What goes up must come down," was about to demonstrate that its validity was not restricted to gravitational forces.

The post-World War II economic "golden age," which both the United States and Western Europe enjoyed, gradually lost its glow as the 1970s advanced. Similar to the general U.S. economy, the carpet industry experienced a marked slowdown in productivity and growth. As Randall L. Patton observed in *Shaw Industries: A History,* "If it had been almost impossible to fail in the 1960s, it became exceedingly difficult to succeed in the carpet business by the end of the 1970s… During the 1970s, the industry went from explosive growth to near bust."

As a consequence, the number of carpet manufacturing companies in the United States declined by approximately 50 percent during the decade. This was the combined result of many mills becoming absorbed by larger, fiscally stronger firms, along with other mills simply being forced to close their doors.

Since the industry had served as such a proud, visible symbol of the new prosperity and the American dream, national government leaders grew concerned. One prominent carpet executive, part of a delegation representing the Carpet and Rug Institute in Washington, D.C., testified at a special hearing before the Housing Banking Committee that this dramatic downturn had resulted from a combination of inflation, rising costs of raw materials, high interest rates, and low consumer confidence.

CMC is Born, Defying the Odds

With these influences at work, the tenuous business climate of 1981 seemed an inopportune time for the debut of a major new entry into the world of tufting. But despite these external factors, the fledgling company that would become known as "CMC" – first the Charles Monroe Company and later, Card-Monroe Corp. – was born.

Its early activity was relatively simple: selling replaceable gauge parts for tufting machines to carpet makers. Before long, however, CMC became involved in the conversion of existing machines. This service had strong appeal to carpet mills since it was much cheaper than investing in new machines during a slowed economy. Then, in June 1982, CMC sold the first machine it had built from scratch, marking its entrance into the tufting machine industry as a competitor to be reckoned with.

This serial number plate appeared on the first tufting machine rebuilt by Charles Monroe Co. in 1981.

Charles "Charlie" Monroe, who founded CMC with his cousin-in-law, Lewis "Lewie" Card, Jr., said the seemingly adverse economic climate actually proved ideal for their company's inception.

"It turned out to be the perfect time for us," Charlie pointed out. "The carpet mills were considering other sources, ways to save money, and they were very open to checking out alternative suppliers. Actually, their openness to this new venture was pretty amazing."

Lewie agreed, pointing out another benefit: "It allowed us to grow with slowness. I can't imagine what it would have been like getting started in the middle of a booming economy. We just rode the wave as the economy began to recover. We were able to start slowly and gradually accelerate with the industry. Then after a while things just took off like a rocket."

The company's 1981 startup date had not been predicated by some overarching strategic plan. "The timing for starting the company was not intentional," Lewie said. "It was just the time that Charlie and I got together and decided to pull the trigger. There was a show in July – the Atlanta Home Furnishings Show – that we wanted to be a part of to launch the business with the carpet industry, so we could start attracting potential customers. We took a spot on the second floor of the building, temporary exhibit space, along with other machinery people and suppliers to the carpet industry. On the floors above were showrooms where carpet manufacturers maintained permanent exhibit space year-round. We thought it would be a good opportunity to establish our presence.

Charles Monroe (left) and Lewis Card, Jr. displayed the handiwork of their new company at a 1981 industry trade show.

"In reality, the economy had probably bottomed out by the time we started and was already slowly beginning to climb back up. It was the start of the Reagan era – President Ronald Reagan had just come into office earlier that year. So it was an ideal time to start building relationships with customers. They were not putting out fires, but were not too busy yet, either. They had time to meet with us, get to know us and learn what we could offer them."

Not Novices to the Industry

The likelihood of CMC's success was bolstered greatly by several factors. First of all, both Charlie and Lewie had years of experience in the industry. They were able to leverage strong relationships with tufters that had been cultivated through previous employment. And the fact that Charlie's father-in-law (Roy) and Lewie's father (Lewis), had become iconic leaders for the industry provided the necessary endorsement for more than getting their feet in the door at the nearby northwest Georgia mills.

Charlie and Lewie had followed two different routes to arrive at their careers in the tufting world. Lewie, of course, had grown up in the industry, although as a youngster he probably did not fully appreciate what remarkable pioneers his father, Lewis, and uncle, Roy, had been as tufting innovators.

"I vividly remember Cobble Brothers on Main Street from the time I was 5-6 years old," he said. "There was the distinct odor of cutting oils in the machine shop, the same smell of oily rags my dad left in a company truck. I also remember going into the office to get scratch pads and pens. That was probably around 1958, when Cobble Brothers moved its facilities to Riverside Drive.

"I was there on Saturdays, always encountering the same smells. My grandfather, Russell Card, lived in a trailer behind the plant on Riverside, working as a watchman."

In 1965, at the age of 13, Lewie started to work at Card & Co. on weekends and during the summer. He assisted around the stockroom, helping to keep track of inventory. He did that for about two years, then in 1967 spent the summer working in the assembly department.

Lewie began working in the sample department in 1970 and remained there until he graduated from high school in 1971. At that time he began to exert his entrepreneurial spirit, breaking his ties with the tufting industry and going into the construction business to build houses, apartments and duplexes. He later bought a lumber yard in St. Elmo, and it was there that Charlie and Lewie had one of their first business interactions. Charlie had contacted him on behalf of Card & Co. to buy lumber for machine crating.

Unlike Lewie, Charlie had not grown up with a pedigree in tufting. He recalled that in the 1950s, his family had only one carpeted room in their house, the living room. "It was woven, and that was all we could afford." At the time he had no understanding of how carpet was produced and, being just a boy, had little interest in learning about what was involved.

The closest Charlie came to having a family legacy related to floor coverings was his maternal grandmother. Prior to her death in 1951, she had taught others the process of making hooked rugs, but that was as near as his family had ever gotten to the carpet industry. He did, however, possess somewhat of a mechanical heritage.

His father, Granville Fox Monroe, was a dentist who had a knack for being able to do virtually anything with his hands. "Once I watched him and a friend as they rebuilt a tractor engine," Charlie recalled. "We were 'weekend farmers' as a hobby, but we did grow our own hay and oats, and raised horses.

Being around farm tractors, implements and trucks, you couldn't help but pick up at least a basic understanding of mechanical things. Of course, they were pretty simple and straight-forward then.

"If you have ever been around hay balers, you know they require a lot of work and attention, especially the baling twine knotter. Now that I think about it, I guess in that way I was working with 'yarn' even back then," he quipped.

The Monroe family lived in the East Brainerd area of Chattanooga. Charlie, his older sister, Kaye, and younger brother, Steve, all graduated from Tyner High School. In keeping with family tradition, each enrolled at Carson-Newman College in Jefferson City, Tennessee. However, after one year Charlie decided to transfer to the newly renamed University of Tennessee at Chattanooga (UTC).

"When I transferred to Chattanooga, they had advertised that you could go to college for one dollar a day – tuition was about $365 a year. It even sounded like a good deal at the time," he said. "I had received a letter from the head basketball coach, Leon Ford, expressing interest in my going out for the team. Although I worked out with the basketball team all the way up to the start of the season, I was enticed away by a job opportunity a family friend had offered. So I traded in my basketball shoes for a box of rubber bands and started delivering the *Chattanooga News-Free Press* in the afternoons and on Sunday mornings. I had a motor route in the Lake Hills and Highway 58 area and was able to pay for my own tuition and books, as well as save up enough money to eventually buy my first car, a new 1971 MGB."

He graduated from UTC in 1972 with a degree in business administration. Ironically, he admitted, "I remember steering away from all the manufacturing courses because I was certain that was not for me. So very little did I know then!"

From College to Tuftco

Charlie had become introduced to the Card family when he started dating Roy's daughter, Renee, while they both were attending UTC. They became engaged to marry, and after he graduated from UTC, Charlie was able to get a job with Card & Co., a subsidiary of Tuftco located on Riverside Drive.

"After graduating I had scouted out the job market and almost decided to take a job with a supplier of checks to the banking industry. But I was very curious about my soon-to-be father-in-law's business. Roy was president of Card & Co. and executive vice president of Tuftco. So I showed up one day and filled out an application, without telling him in advance. I don't know if this was a surprise or not.

"Roy's office manager, Mildred Brown, at the time was not certain that I would like the tufting machinery business, so she tried to schedule several job interviews for me with their suppliers. Finally, Wallace Hammel, the manufacturing manager, decided to hire me. He offered one piece of advice for me as the future son-in-law of Roy Card: 'Family members are expected to set the example, not be the exception.' I will never forget those words – they still ring in my ears today."

Hammel was a good person for Charlie to learn from in other respects as well, since he had been involved in the industry since the early days of creating the scroll machine, and was one of the developers of clutch and roll scroll and the Tuftco Yarn Control Systems, working with Lewis and Roy. He was generally regarded as an "electronics guru."

The Card & Co. subsidiary where Charlie went to work made both tufting machines and pattern attachments for two Tuftco machines – the Card frame, which had a cast iron head (top housing) and bed plate, and the Southern frame, which was fabricated from steel plate (originally developed by Southern Machine Company that had been enfolded into Tuftco). With Card & Co., Charlie received a thorough, firsthand introduction to all areas of the business – experience that would serve well in preparing him for the leadership role he would assume with CMC years later.

First he worked in the plant, in the head and bed department, where main shafts are installed into the head and bed plates (internal parts of the tufting machine). "There I learned the difference between a drill and a tap," he laughed, noting that for veterans of the industry the distinction is so obvious it does not even bear mentioning. (A drill makes a hole; a tap cuts threads so a bolt can screw within the threads in the hole.)

Next he moved to the assembly department, where motors, cloth feeds and yarn feeds all are attached to the main frame of the machines. After about two months, Charlie proceeded to the sample department, where tufting machines with 42-inch sewing widths were used to produce samples for customers. He learned the practical applications of the tufting machine and some of the sewing properties and capabilities of different machines.

In that department, machines are continuously converted from one type of pile to another to accommodate customers' needs to see samples of what can be made on their specific machines. For instance, a conversion might require moving from 3/8-gauge cut pile shag machine to cut-loop shag, or from 3/8-gauge shag to 5/16-gauge shag. These conversions required changing gauge parts and rethreading the machines for the new gauges.

"This gave me an understanding of the function of the yarn feed, cloth feed, and how pattern attachments were able to create different types of carpet constructions and its variations," Charlie noted. Overall he was in those areas for five or six months.

Next he was asked to work on costing inventory of manufactured parts, determining an accurate cost for every part that was made. This assignment led up to the year-end inventory, and since Tuftco was a public company that was audited by an outside accounting firm, "inventories were a big deal," he stated. Performing this kind of inventory was a difficult task because it was a time of rapid inflation, and computers had not yet arrived to simplify the calculation process.

Because of the extreme inflation during those years, the company made a change from a FIFO (first in, first out) method of inventory accounting to LIFO (last in, first out), according to Charlie. All inventory had to be priced twice, once for the base year cost and once for the current year cost. These calculations had to be done by hand, since computers were not yet being utilized by smaller companies. So he not only learned how the machines functioned and what they could do, but also what was involved in manufacturing the myriad parts that were required to build such increasingly sophisticated machinery.

This process of working through the different departments of Card & Co., learning the intricacies of the various areas of the industry, amounted to a management training program. However, Charlie said he was never aware of any predetermined intent to prepare him for any given job in the company. "If there was any grand scheme of things, Roy never told me!"

After being in charge of inventory, he then went on to become purchasing agent for Card & Co., a role he held for about 2½ years, from 1973 to 1975. His time in the various departments prior to becoming purchasing agent spanned approximately 10 months. While Charlie was involved in purchasing, he took a night class at Kirkman Vocational School. This machine shop course taught him how to operate a lathe and provided him with a hands-on feel for what goes on in a commercial machine shop, giving him a better appreciation and understanding of the machining processes.

Moving to the Sales Side

Then in 1975, Roy approached Charlie and told him they were combining the Southern and Card & Co. sales forces, and he wanted Charlie to go into sales. This involved selling both the Southern and Card machines and working out a plan for how the merging of sales teams could be done most effectively. So Charlie reported for work at the Card & Co. site on Riverside Drive, but also maintained an office at the Tuftco corporate facility on Holtzclaw Avenue.

He began the sales portion of his still young career by going to the July 1975 carpet industry market in Atlanta with Bob Humphreys, Tuftco's vice president of sales, and Bill Fowler, technical sales representative. That represented the first attempt to form a common sales force between Card and Southern. Until then, the sales teams and plants had competed against each other, promoting their respective machines over the other division's product.

Relationships proved very important as they proceeded, since mills tended to favor one machine over the other, partly due to strong established ties with both management and sales representatives, and familiarity with one frame or the other. Ultimately the effort succeeded, and in 1979 a decision was made to merge the manufacturing plants. Card & Co. was moved from Riverside Drive to an expanded plant on Holtzclaw Avenue, resulting in bringing the production of two distinct machine types under the same roof.

A 'Homecoming' for Lewie

Beginning in late 1972 and early 1973, Lewie had taken on a special assignment to move to Sanford, Florida for an interim period, overseeing Tuftco Homes, a mobile home manufacturing plant that his father and Jack Frost had initiated. The company also maintained two other plants in Alabama.

Once his responsibilities there had been fulfilled, Lewie returned to the tufting machine business in January 1975, setting up an office at Tuftco soon after his mother, Katherine, had passed away. He became a purchasing agent at Card & Co. in July of that year, replacing Charlie, who had gone into sales.

For most of the next two years Lewie and Charlie settled into their respective roles, but again changes loomed on the horizon. Toward the end of 1976, both Lewis and Roy decided to sell Tuftco, feeling that their entrepreneurial instincts were being stifled. And at the start of 1977, Tuftco was sold to a New York-based holding company, Dyson-Kisner. As is often the case in an acquisition, management philosophies and the corporate culture began to undergo marked changes.

Lewis and Roy remained on the Tuftco leadership team for several years, but longed for the day when they could regain their independence as innovators. In the meantime, son Lewie also became receptive to new opportunities, especially since he no longer had a financial interest in Tuftco.

One such opportunity came about when Rodney Hill, vice president of Card & Co., and Gerald Kaiser, the service manager, approached him about helping to finance the startup of Pro Tufters, a commission tufting business.

"They talked me into going into the business with them – they came to me, I didn't go to them," Lewie stated. "It was one of the hardest days of my life telling Roy what I had done, taking his two right-hand men away from him. Actually, I never really left Card & Co. – I was only an investor with Hill and Kaiser."

Despite not being the initiator of this new venture, for some time Lewie felt like a bit of an outcast for being involved in the exodus of two such valuable Tuftco employees.

Thinking back on that event, Charlie agreed. "I remember the Saturday morning after the news came down. Roy was in the process of building a new home on the lake and I watched him sweeping out the construction debris. I'll never forget the look on his face – concerned, angry, hurt. You could see him processing what had just happened."

That was not the only processing that was necessary. Pro Tufters was located on Cleveland Highway in Dalton, and although Hill and Kaiser had worked together well at Card & Co., "as business partners they proved to be like oil and water – they didn't mix well at all. They were two very strong-willed people," Lewie remembered.

Roy remained upset with Lewie for quite a while, although Lewie did give Tuftco an order for two tufting machines – costing about $500,000 at the time – during the relatively brief time Pro Tufters was in operation.

Assuming a Broader Operational Role

The silver lining was that Lewie's partnership with Hill and Kaiser enabled Charlie to return to Card & Co., still a Tuftco subsidiary, and assume a more substantial role in the overall operation.

"There was a union at Southern Machine Co., but Card & Co. had no union, largely because of Roy's fairness with all of the employees. He always fought off union elections (there were four attempts in five years). If you were to ask anybody about Roy Card, they all would say he was unquestionably fair and honest," Charlie said.

"The management at Tuftco was concerned there was always a possibility that Card & Co. might lose the next union election, resulting in having two separate unions – one at Southern on Holtzclaw and the other at Card on Riverside. They knew it would be virtually impossible to merge or combine the plant operations that had two different unions, separate leaders and differing contract dates. So they decided to consolidate the two divisions, for manufacturing efficiency and to avoid a potential union conflict. They converted Card & Co. from a subsidiary into a division of Tuftco, and Card assumed Southern Machine's contract with the machinists' union.

Ironically, after successfully fending off the union for years, Card & Co. allowed its employees to join the union if they chose to do that. Card & Co., now a division of Tuftco, in effect was stripped of its autonomy. As a subsidiary, it had been its own corporate entity, with separate officers and board, and financial responsibility. It was owned by Tuftco, but had retained its own corporate charter.

As Charlie returned to Card & Co., he continued in sales but also was put in charge of the sample department. Soon afterward, in 1979, most of the Card & Co. operation was moved to the Holtzclaw location, with the exception of the sample department and the machine technology and development department, which remained on Riverside Drive.

During the time of the moving process, Roy Card conceived the idea for inventing the "Graphics" machine," an extremely significant development that eventually would become another foundational patented advancement with worldwide significance for the industry. This innovation utilized double-shifting needle bars and unique threading sequences to create an all-new array of geometric patterns.

Charlie commented, "When I saw the first pattern, I marveled at how beautiful it was. Roy simply said, 'This is only the beginning. You have not seen anything yet.' Development of the Hydrashift helped to advance this machine.

"I remember sitting in the office with Roy and Lewis, and they asked me what we should call the machine. 'We should call it the Graphics machine,' I replied, because of its amazing design capabilities, and they agreed. It was the first device in the industry that I had the opportunity to name."

Keeping Wheels of Progress Moving

"One day Jack Frost called me, asking when we would be moving the sample department to the Holtzclaw plant. 'It looks like I'll have to come over and move it myself,' he said. We were in the midst of developing the Graphics machine, running sample patterns and experimenting with what it could do. If we would have stopped and taken the time necessary to move the department – tearing down and reassembling five or six machines – it would have stalled the development of the Graphics machine, so I simply deferred to Roy, my superior," Charlie said.

"We knew we were onto something huge for the industry and for Tuftco, so we delayed moving until we had the Graphics machine to a point where we could relocate without impeding progress. We started moving 2-3 months after Jack had called me, and by then we had worked out the kinks and had sold several machines.

"Understandably, Jack's background was in accounting and administration – not mechanics – and his interest was in expediting and completing the move to merge the two operations. At the time, Southern Machine had three sample machines, but no pattern machines. The service technicians operated their machines, but there was no significant development of new technology going on there. Card & Co. was where most machine development took place. We felt it was important to keep working on the development of the Graphics machine, especially since it was at such a critical stage."

After the Card & Co. sample department was moved to the now-combined Tuftco plant, Charlie headed the new sample and machinery development department, while retaining his sales responsibilities.

"This was still before the days of computer design," he noted, "and having learned the Graphics machine from the ground up, I would use two transparent sheets of graph paper and colored pencils to manually illustrate patterns it could produce. I would chart the sequences – graphing the movements and threading sequence for one needle bar on one sheet, then the movements and threading sequence for the other needle bar on the second sheet, to show how they worked together to produce a resulting pattern.

"After that I began to teach other people, primarily customers and designers, how to do the same thing. One day computers would be able to do that far more easily and efficiently, but my knowledge of the Graphics machine and the process of graphing the designs actually became very significant to my role in starting CMC."

More and More Ready for Change

Meanwhile the seeds of unrest were sprouting in the hearts of the Card brothers, causing them to yearn all the more for the days when they could operate independently again, having greater freedom and latitude to express their entrepreneurial instincts.

"In 1979 we all were at Tuftco on Holtzclaw," Charlie recalled. "Lewie had left to go into the mobile home park business. Roy and Lewis were in the third and final year of their management and employment contracts with Tuftco.

"Every day I could see on Roy's face the frustration that the challenges of two operations and two different management approaches and cultures were having on him. Often I would hear both Roy and Lewis say, 'I can't get anything done.' Lewis was chairman, Jack was president and Roy was the executive vice president, but nobody knew where to look for leadership because of the conflicting philosophies."

Tuftco still was selling two distinctively different tufting machines – the

Southern machine and the Card machine – and often salespeople would push the one they were most familiar with. That was true for Charlie as well, who was having great success with the Graphics machine.

"We had begun selling to Shaw Industries in 1975, and I had developed a good relationship with them. I remember going down to Dalton one day and they wanted us to quote on two tufting machines. I asked if they wanted the Southern or Card machines. They replied, 'We don't know – we want the best.' So having the most confidence in the Card machines, I sold them two of those.

"Shaw was on a tremendous growth track at the time, and I had built a strong relationship with Norris Little, their vice president of operations. We had begun selling the Graphics machine in December 1979 and early 1980, and Shaw had an option to purchase one of the first ones that we built, at an introductory price. But by the end of 1980, they decided they did not want that machine.

"Since I had experience with the Graphics machine and its patterning capabilities, I asked Lewis and Roy for advice about going into the commercial tufting business. The idea was to call the company Card-Monroe Carpet. I had several talks with them about the pros and cons of commercial tufting. Roy had been my mentor in all of our discussions about commercial tufting with the Graphics machine. I even approached Norris Little about selling me the tufting machine on which they had the first option – they were not involved in manufacturing commercial carpet at the time, and the machine did not fit with what they were doing.

"Eventually we elected not to do that, but Roy and Lewis packed up their desks and left Tuftco at the end of 1980. Their employment contracts had been fulfilled, but they still had a portion of their non-compete agreement to honor."

The brothers did not sit idle, however. They redirected their focus onto managing their investments along with doing some independent developmental work. Ultimately, their careers and the tufting industry would converge again, allowing them to continue exploring the frontiers of machine innovation.

Going Overseas – and Then on His Own

In January 1981 Charlie had made his first business trip overseas, representing Tuftco at the Domotex market and visiting the Dura Carpet Mill in Fulda, Germany. He took a Programmable Read-Only Memory (PROM) machine with him, to teach the customer, Dura, how to program the PROM modules to achieve different needle bar stepping movements, especially for the Graphics machine. The needle bars were driven laterally by an electro-hydraulically controlled cylinder, an attachment called the "Hydrashift."

"Roy and Lewis had already left Tuftco by that time, and Jack gave me a nice increase in compensation, which I appreciated. But I became very uncomfortable with continuing to work there – it just did not feel right. Card & Co. had been much smaller, and felt more entrepreneurial and progressive. When we all got pulled together into Tuftco, it just didn't seem the same. I enjoyed what I was doing, selling and being involved with machine development, but still felt dissatisfied, so I decided to try going on my own."

Back in the summer of 1979, Lewis and Lewie had established an office on Adams Road in the Hixson area of Chattanooga for overseeing family businesses. "My dad and I had a 32-by-64-foot building out here, along with a 5,000-square-foot warehouse for our 'junk.' We had acquired the property from a developer of the adjacent Racquet Club, a private tennis club that we had invested in," Lewie explained.

"When we got started there were just four of us – Dad and me, a bookkeeper and an accountant. Then Roy came, and Lamar Wheat, a family friend in the carpet business, had a little office on the other side of the building. That was the place where it would all begin – a little office, some warehouse space and a bit of land, about five acres at the time."

After Charlie had reached the conclusion it was time for him to leave Tuftco, he went to see Lewie at his Hixson office. "I felt that since I had been involved in creating carpet patterns for so many people, if nothing else I could fall back on those relationships and sell them hooks, knives and needles while I was continuing to do carpet patterns. Lewie and I decided to put together a business to produce spare parts and do machinery conversions, eventually also rebuilding tufting machines," he said.

"Before I announced my plans to leave Tuftco, I had a talk with Edgar Pickering, a lifelong friend of Lewis and Roy – especially Lewis. Edgar had managed the plant in Blackburn, England for Cobble, and was then selling machines for Tuftco in Europe.

"Just days before I would approach Jack Frost and explain my intention, I told Edgar I was going to leave Tuftco to go into business for myself. However, I asked him, 'Don't act like you know until I tell Jack first.' I remember Edgar responding, 'I don't want to have any part of deception.' I probably learned more from that statement than the many things I had learned in business school. It impressed upon me the reality that I needed to think about what I was doing, how I was doing it, and what I would be asking others to do. That's an important lesson that I have had to apply many times since then.

"When I met with Jack, he said I might as well go ahead and leave immediately. I asked if I could sell Tuftco machines for him on a commission basis. Jack said he would have to think about it and get back with me, but later he said no. So I left and established an office on Adams Road with Lewie on April 1, 1981, where our CMC plant currently stands."

In Business, Selling Gauge Parts

Charlie's first step in starting the Charles Monroe Company was to align with Erwin Enterprises to sell Eisbar needles and Schlemper knives for tufting machines. He also connected with Larry Gable, who then had his own company, Metal Crafters, to make loopers for him. The loopers would be sold under the CMC label. The business was officially incorporated on June 1, 1981, but the possibility of manufacturing tufting machines was "so far away from reality at the time – we had to take things one step at a time then," Charlie said.

Since the carpet industry had been undergoing major consolidation, both he and Lewie felt some trepidation but never debated whether they made the right decision. "(Consolidation) started about the same time we were going into business, with Shaw taking the lead in buying companies. Then Mohawk and others began following suit. At first it gave us a lot of concerns, since we didn't know what to expect as the field of potential customers began to shrink. But it probably ended up being a real benefit for us.

"As we built relationships, we were able to establish trust and demonstrate that we would provide quality equipment. And when we eventually got into manufacturing new machines, ours became the machine of choice for Shaw as well as for the majority of North American carpet manufacturers."

From its first month, CMC was profitable, selling hooks (loopers), knives and needles for tufting machines. "Eventually, Tuftco allowed me to buy a few parts from them, which I then resold." In addition, Charlie's company was creating graphics patterns. He also bought knife holders from Lester Cobble of Cobble Machinery Co., which was not associated with the Cobble division of Spencer Wright Industries (the former Cobble Brothers Machinery Company).

"Bentley Carpet Mills in Los Angeles was one of the companies that relied on me for patterns. As a result, they would buy all of their hooks, knives and needles from us, and eventually tufting machines," he said.

"Albert Cobble, with his grandson, Buddy Cobble, had some manufacturing equipment for gauge parts in east Chattanooga. We went to negotiate with Albert to buy the equipment. He was a tough negotiator even then, well into his 80's. Eventually we worked out a deal and acquired some used metal-working equipment, along with some fixtures, got it all onto a truck and brought it over here, putting it into our warehouse. We used some of it, but it never really gave us the results we wanted."

Key Ingredient Added to the Mix

Not long after opening, Charles Monroe Company made a strategic alliance that has continued to pay dividends into the present.

"Roy and Lewis told us about James Jackson, who had a slat pattern manufacturing operation in Chickamauga, Georgia, and also made other spare parts for the tufting industry," Charlie said. "When we had still been with Tuftco, Milliken and Co. Carpet Division had challenged us to improve the accuracy of the gauge parts – needle bars and hook bars. Years later, when we finally accomplished that at CMC, we actually asked Milliken to test the accuracy of our gauge parts since they had one of the most sophisticated labs with the measuring equipment available for doing that test. We used the letter they wrote reporting their findings as a sales tool in working with other customers.

"Jackson had created gauge part manufacturing equipment when he worked with Lewis and Roy at Cobble, and then again at Card & Company they had hired James to make their gauge parts equipment. This was not the kind of thing you could just go out and buy at a local hardware store. It was very specialized tooling for adapting milling machines to be able to manufacture gauge parts to strict requirements of very close tolerances. We knew that if we were to meet Milliken's criteria, James Jackson would be the one to do it," Charlie pointed out.

"So we went to see him and his son, Sonny, at their little machining operation in Chickamauga. We told them what we wanted to do and that we desired for them to be a part of it. To close the deal, we traded them stock for their assets and James Jackson & Co. ceased to be, becoming instead a part of CMC. We couldn't afford to buy him, but could afford to have him merge with us and receive stock in our company in exchange.

"Before that we had bought our parts from a variety of other sources that had differing degrees of expertise in the manufacture of gauge parts. Finally we had a top-quality machinist and developer for manufacturing gauge parts. With James we were able to start producing needle bars, needle plates, hook bars, knife holders and looper bars, all in house.'"

The advantages of merging Jackson's company with CMC became immediately apparent, according to Charlie. "All of a sudden we could make our own gauge parts. John O'Rear, who had been the machine shop foreman at Cobble for many years, came to help set up our machine shop. And Morris

Lovelady, who had worked with Card & Co. since 1972, joined us as our engineer and head of production control.

"With people like this on board, we were able to start converting and rebuilding tufting machines. We had purchased some secondhand machines so we could rebuild and refurbish secondhand frames. We bought Card machine frames to sell as converted machines. We used only the frame in remanufactured machines, putting in new internal parts and gauge parts to meet our customers' requirements."

Always a Technician, Never a Tufter

Jackson, who died in 2004 at the age of 77, was a true specialist. He brought to CMC the capacity to make the precision parts necessary for its increasingly complex tufting machines, but never got very involved in the technical aspects of how to make the machines operate with greatest efficiency. That challenge he left to Lewis, Roy and others. His focus was on the accuracy of the parts themselves which, in turn, made it possible for the machines to run more efficiently and to make better quality carpet.

With his son, Sonny, he had started Jackson Machine Company in 1971, specializing in the manufacture of gauge parts. For him, it was more than a craft – it was his passion.

"To most people he was a fairly private person. He talked a lot about going fishing and playing golf, but he spent most of his time trying to better himself," Sonny recalled. "Everything in the machine shop he thoroughly understood; he knew how to run every machine before anyone else did.

"Daddy worked all the time, and if he wasn't studying up on how to improve himself, he would have a sideline business going, he and his brother, Rex. Daddy ran the machine shop, and Uncle Rex specialized in sheet metal. Daddy working – that's all I can remember as a kid. That left very little time for what we now call 'hanging out,' father and son. Between 1971 and 1981, when we got into business together, we actually got to spend more time together in recreational activities – fishing and hunting.

"But working was what he loved to do. I heard him talking all the time about parts he was working on. When CMC started, the way we made needle bars, knife blocks and cut hook bars, each required a specific, handcrafted piece of equipment to put them together. Many times these had to be fashioned out of multiple types of equipment to create machines for a specific job. Over time we built many turnkey machines for making specialized parts – our task was to design the machine to produce a specific part in the most cost-effective way."

After Roy and Lewis Card, Sr. joined CMC at the beginning of 1983, having fulfilled their non-compete obligation with Tuftco, Lewis and Jackson established a particularly effective working synergy, with Jackson given the task of figuring out how to make parts work for different applications.

Lewis and Roy Card stand with Charles Monroe and Lewie Card soon after joining the leadership team at Card-Monroe Corp. in 1983.

Sonny stated, "One time I heard Lewis ask him, 'How can we make a needle bar drilling machine that will guarantee the accuracy we're promising?' referring to a project they were working on for Milliken. After a lot of hard work, Lewis and Daddy were able to deliver to Milliken a needle bar they had manufactured, guaranteeing that it would have a total accumulative error tolerance within 5/1,000 of an inch over the needle bar's 15-foot span from one end to the other. Milliken's lab responded that they had found the bar to be accurate within 2/1,000ths, and called it one of the best bars they had ever seen."

That demonstration, the younger Jackson noted, was "instrumental in our getting Milliken's business and we still have it today. It was very impressive for something to be that close (in accuracy) in those days. Prior to CMC's System One cutting system, none of the relationships between needles, hooks and knives had to be that close, since they were all adjustable."

For machines that were delivered to customers before such precision could be achieved, service technicians sometimes spent several days making adjustments to properly align all of the gauge parts. However, these time-consuming adjustments were unnecessary once CMC was able to manufacture its highly accurate gauge parts.

"It was a new paradigm – we were met with a lot of resistance to the idea that it would even work," Sonny Jackson noted. "The technician's concern was, 'You're taking away what I do. I go in and make it sew.' But it was a real selling point to our customers. Suddenly we had taken much of the human element out of it and put it into the accuracy of the machinery, making it possible to easily grab a knife block off a shelf and fix or replace a worn part in seconds. It became very operator-friendly compared to what it had been in the past."

In the Spirit of Competition

At the beginning of 1983, the company's name changed from Charles Monroe Company to Card-Monroe Corp. when Lewis and Roy invested in it, becoming stockholders and officers in the company, as chairman and vice chairman, respectively.

Having his father and uncle become part of the business provided an opportunity for Lewie to team with Charlie in the everyday operation of the business, while the brothers resumed their focus on development and innovation, continuing to capitalize on their respective strengths. Lewis's forte remained mechanical development, working to improve cutting systems and increase machine speeds. Roy focused his expertise on the technical side, making changes that would make it possible to take an idea he had for a new type of fabric and develop it into reality.

Roy Card, Charles Monroe and Lewis Card, Sr. are shown with Jon Shaheen, among CMC's first customers for new machines.

"Someone once had a great description of what their differences looked like," Lewie recalled. "Roy always had a piece of graph paper in front of him (for creating new fabric designs), while Dad always had a machine part in his hand, wondering how it could be made better.

"It was a wonderful opportunity to have had them working with us,

143

available to help whenever they were needed," Lewie said. "Any time we had questions, we could go to them.

"Even today we still have had some of that opportunity. They don't get involved in day-to-day operations, but are always willing to advise us when necessary."

Within the brief span of three years, building on an industrial legacy that had been growing since the days prior to World War II, CMC had made its presence known in the world of tufting machine manufacturing. Suddenly the established entities in the industry – Tuftco and Spencer Wright Industries (Cobble) – found themselves with a new, formidable competitor.

"Bob Harmon of Richmond Carpet Mills in Ringgold, Ga. became one of our first customers of rebuilt machines," Charlie recalled. "Years later, Bob and several of us at CMC were reminiscing about the early days of our company, and we thanked him for taking a risk with us. He jokingly responded, 'No one else would give me credit!' although that was definitely not true."

Other early CMC customers included Collins & Aikman Carpet Mill in Dalton; Wellco, a division of Mannington in Calhoun; and Callaway Tufters in Dalton.

June 7, 1982 became a milestone date for the company, marking the completion of the first new CMC machine, which was sold to John Burnes, vice president of manufacturing for Marglen in Rome, Ga. Another of the initial customers for new CMC machines was Alfombras Mohawk in Mexico.

A rebuilt tufting machine produced by Charles Monroe Company in 1981.

Workers who combined their skills to produce the first CMC machine in 1981 were, top (from left), Robert Cameron, Ken Stephenson, Larry Pickens, Chuck Gilbert, Aaron Simmons, Herb Corsey and Wendell Snyder; and in front, Morris Lovelady, Mike Shipley, Robert Pelfrey, Wayne Williams, Mac MacGregor, Jim Higdon and John O'Rear.

Max Beasley, who had participated in the technological advances of the tufting machine industry for many years, observing numerous corporate comings and goings, described the impact of CMC's arrival onto the scene:

"When CMC began their business, there started the fiercest period of competition I have ever seen. ... This had to be a good thing for the tufting industry. While there was always a push to get tufting into other textile products and to make tufting machines more efficient, there was nothing like competition to speed things up! ... I would guess there has been at least twice as much progress made in this field in the last 20 years as there had been in the previous 30 years – or since the tufting industry began.

"No matter who made an innovation, the others would have to react to it – and quickly, too. One of the first ideas that came out was a new kind of knife holder. It was a really ingenious design that embodied the principles Lewis and I had worked with, holding a whole inch of knives in one knife block...

"Then CMC came out with a new way of mounting the knife block in the machine, again striving to make it so that it could not be installed any way but the most efficient and accurate way. ... (Tuftco) had to come up with our own version that would have somewhat of the same result."

Innovation became an integral part of each company's success. Since tufting machines were so well made and durable, they were regarded as "virtually indestructible." Enhanced technology and improved carpet-making capabilities were the keys to marketing new machines.

As Lewie Card observed, "Every generation was hungry to make a mark in the industry. There is always a place for someone good to make improvements or enhancements."

At the same time, since demand was limited for such highly specialized machines, it was generally understood that the industry could only accommodate a handful of competitors, he said. "I remember Spencer Wright's right-hand man, Bill McCamy, telling my father years ago, in the early '80s, 'There is not room for more than two (tufting machine companies) in this industry.' To which my father replied, 'You're right, and we're going to be one of them.'"

Getting in on the Ground Floor

Today, Allen Neely serves as plant manager at CMC, having been with the company for about 28 years. He arrived in 1981, two months after the hiring process began at then-new CMC, having spent six months with Cobble as a machinist.

Neely started as one of five 18-to-20-year-old machinists in the production department. "One day Lewis came through the shop and challenged each of us on a small project basis, making gauge parts. After I was here about two years, Lewis came to me and said he had selected me to work in the small R&D department we maintained in a small production facility. Initially, we had only 10,000 square feet for our manufacturing, production and service work. At that time I had little interaction with either Charlie Monroe or Lewie Card. I worked primarily for Lewis."

He has retained many vivid memories of that time. "I worked with Lewis probably about 10 years until he retired from active work. We were project-driven, and he was relentless. We worked six days a week, 10 hours a day. I worked many Saturdays when it was just him and me here – no one else around. He was always here with the exception of a six-week vacation period he always took the middle of each winter in Sarasota, Florida. During that time he would appoint me to be the leader of the small R&D machine shop effort.

"Lewis set the highest work ethic standard of anybody, always professional and always business. He had little time for building relationships, until the years after his retirement."

In the late 1980s or early 1990s, as Lewis entered semi-retirement, Neely began to strengthen his working relationship with Roy. "Up to that time, Roy had terrified me. Lewis had always handled the mechanical side of development, and Roy had the carpet or fabric side. Roy knew what he wanted in carpets, and Lewis understood the mechanics that could produce those effects and objectives.

Some of CMC's department managers in 1995 included (from left)
Bill Ristom, John O'Rear, Morris Lovelady, Lawrence Spurling, Bill
Christman, Danny Henderson, Allen Neely and Johnny Caldwell.

"My initial thoughts about Roy had been conditioned by perceptions
other people had of him. Lewis was very liberal with money and machinery,
while Roy was very conservative. And Lewis was much more of a risk-taker
than Roy. I had been warned that you don't ask Roy for anything, because
you were not going to get it – but that truly was not what I experienced over
my years of working with him."

Shoring up the Sales Team

Jerry Hendricks, who had been in the tufting business since 1958, became
another key addition to the CMC team in 1985, resuming his longstanding
relationship with Roy Card that had included jobs with Cobble Brothers and
Singer-Cobble.

After starting in the tool room and machine shop with Super Tufter,
and then its shipping and receiving department, Hendricks had moved to
Cobble in 1960. "Roy was looking for some younger technicians to come on
board, and I asked him to be considered. He came back to me and gave me
a chance to be a service technician. Technicians always started in the sample
department, learning the groundwork of the tufting machine and its set-up.
Several different people over the sample department helped me to learn the
ropes."

Around 1965, Hendricks went out on the road for the first time, spending
about 10 years "sewing off" machines, installing various attachments, and
trouble-shooting for machines needing repairs. He traveled across the United
States ("from Massachusetts to California – we did a lot of upholstery fabric

machines back then"), as well as to New Zealand, Japan, Canada and Australia.

In the mid-70's he took a break from traveling and became a service manager with Singer-Cobble, a role he filled for about three years. Then he accepted an opportunity to relocate to California, where he established an office for taking care of West Coast customers. "The plan was to do both sales and service, but after I got out there, I discovered it was almost all service – trouble-shooting, basically." He did that for about 18 months before returning to Chattanooga.

Back in Chattanooga, Hendricks was placed over the sample department – the "textile lab" – overseeing about six people working in that department. They performed demonstrations for customers wanting to look at the effects of different yarns or gauges, and helped to develop some tufting machine refinements. "In the late 70's, we put the first pieces of the new LCL (level cut loop) machine together."

In 1981-82, following a layoff at Cobble, Hendricks operated his own business for about a year, rebuilding and refurbishing machines with Wayne Kite and Earl Suggs. He then took a job with the Greenwood Corporation, which manufactured continuous dyeing equipment in Dalton, an advancement from beck or piece-dyeing.

It was at a carpet show in Atlanta, where Greenwood and CMC were both participating, when Hendricks "got to talking" with Harold North and was asked if he was interested in joining the CMC sales staff. He was – and he did. For a time, he served as interim service manager, but has spent most of his time at CMC as a sales executive. "I have always enjoyed both the sales and technical sides, but being in sales lets you get involved with customers and build relationships with them. You feel like you're part of helping them grow their business by getting them into another type of machinery or product."

He cited the example of Dalton's Clayton Miller Hospitality, one of CMC's first customers to use the computer-controlled, servo-motor driven Infinity machine. "It was a huge step for him to move into that arena as a smaller carpet manufacturer, but because of his willingness to venture into the unknown, he is now recognized as one of the best in the hospitality business."

The company's main focus is hotels, motels and ballrooms, making carpet with very specialized designs – customizing colors, patterns and logos for each individual customer. "That separated Clayton Miller from other players early on – he was willing to do that."

Hendricks's longevity in the business has given him a full appreciation of the phenomenal impact that computers have had on the tufting industry.

Starting in the late 1970's to early '80s, CAD (computer-aided design) systems became part of every aspect of engineering design.

CAD systems offered the ability to view drawings on computer, resulting in many improvements being made and a new era being born for carpet development. The systems made it easier to create new patterns and effects, and the designers could see the results much faster. Design engineers would create a visual image of the design even before it was produced on the machine.

"They could see pattern lines on-screen, helping them to determine how it would look on the carpet," Hendricks commented, acknowledging how much the industry has progressed over the five decades he has been a part of it.

That was merely the start. In the succeeding years, computer-enhanced design and computer-driven machines would have an amazingly transforming impact on carpet manufacture. At the same time, CMC leadership would recognize the importance of formulating clear-cut statements of mission and values to serve as a compass for the company in dealing with the challenges of a fluctuating economy, continued growth and innovation, and global expansion.

Charles Monroe (far left) and Lewie Card (far right) are shown with CMC employees John O'Rear, Mike Shipley, Aaron Simmons, Wayne Williams, Chuck Gilbert, Herb Corsey, Jim Higdon, Morris Lovelady and Mac MacGregor.

11

CARRYING ON THE LEGACY

It's unlikely that anyone who has made a meaningful, positive impact on society, whether as an explorer, artist, statesman, scientist, industrialist, educator or inventor, started out with the declared objective, "I'm going to establish a legacy for myself." Legacies come about as the byproduct of lifetimes of hard work and initiative – memorials to the vision, determination and talent that led to noteworthy achievement.

This may be why it seems difficult for someone like Lewis Card, Sr. to regard himself in terms of a rich legacy, even though that is what his life has produced. His career started in 1939 – at a time when a young man was grateful just to have a job, any kind of job – spanned the remainder of the 20th century, and stretched into the 21st century. In 2008, more than 100 years after young Catherine Evans Whitener first laid her eyes on a primitive, unusually decorated bedspread, Lewis reflected on the long and sometimes tortuous path the craft and technology of tufting have traveled.

"I never could have envisioned what would one day happen in the industry. I didn't have that much imagination. I figured that if we could make a throw rug machine, we could somehow make a carpet machine, but that was about it."

He and his brother, Roy, were both so preoccupied with the challenges of each day, they rarely had time – or the inclination – to reflect on what they had accomplished or where their industry might be headed. Each day, it seemed, had enough trouble of its own.

"Our toughest customers were the ones in the carpet weaving business. Most people in the weaving mills would say that tufting machines could never make carpet. They couldn't learn or couldn't see – or didn't want to see – the potential for tufting machines. They didn't want the machines, but eventually they were forced to buy them because everyone else in the carpet industry was getting into tufting, the growing part of the carpet market."

Although Lewis has been effectively retired for more than a decade, he and Roy have maintained a degree of involvement in the carpet tufting machine industry by serving on CMC's family-governed board of directors. Roy's presence is still being felt in the company as he continues to work several days a week, still drawing from his nearly 60 years of experience to offer input on how the CMC machines can be improved.

Reflecting on his career when it was at its heyday, Roy commented, "We forever had people bringing woven samples to us, saying they wanted to make the same things on their tufting machine – so that they could produce it ten times faster. And a great majority of those things we eventually were able to do. But the things that we have seen in the last five or six years, it just amazes you what can be done now with tufting machines. They now have pattern capabilities and computer mechanisms that we never imagined.

"I guess I was 20-30 years too soon. I could never learn enough about computers today. I know what I would want them to do, but wouldn't know how (using a computer) to make the machine do it."

The Card brothers' legacy, of course, was solidified long ago. They have stayed close to the business simply because the passage of years has not cooled their passion for the work, and have retained the curiosity to learn the next, greatest thing that their machines can do.

Meanwhile, Lewie Card, Jr. and Charlie Monroe, the family's third generation in the industry, are doing their part to ensure and extend the Cobble-Card legacy. They have learned what it means to stand on the shoulders of giants.

From Start-up to Pacesetter

As with the so-called experts that once dismissed tufted carpet as a passing fad, anyone who might have questioned CMC's capacity to survive in a highly specialized industry has been forced to admit the folly of their skepticism. More than 25 years later, CMC – under the direction of Charlie and Lewie – now holds a prominent place among tufting machine manufacturers.

Since its inception in 1981, CMC has built more than 1,800 tufting machines. These are being operated not only in mills across the United States, but also around the world. Through mid-2008, CMC machines also had been sold to carpet companies in Argentina, Australia, Belgium, Brazil, Canada, China, Czech Republic, Denmark, England, France, Germany, India, Iran, Iraq, Ireland, Italy, Japan, Mexico, the Netherlands, New Zealand, Northern Ireland, Portugal, Russia, Saudi Arabia, Scotland, South Africa,

South Korea, Spain, Taiwan, Thailand, Tunisia and Turkey. In 2005, one of CMC's customers in China reported they had produced more than 5.5 million square yards (4.6 million square meters) of carpet over a six-year period for Chinese consumers.

The underlying manufacturing principles remain unaltered: needles, yarn, backing material, hooks and knives all working together to create a visually appealing and functional flooring fabric. For every other aspect of the industry, however, the one constant seems to be a state of continual change and improvement. As Lewis observed, in the early days it would have been impossible to anticipate or imagine the huge strides the industry would make over the succeeding decades.

Technology, of course, has played a dominant role. Some of today's tufting machines run at speeds approaching or surpassing 2,000 revolutions per minute. Even with that quicker-than-the-eye velocity, as many as 2,600 needles across the width of a five-meter machine achieve superb precision, meshing their corresponding hooks and needles to execute the prescribed patterns, textures and designs. This combination of speed and accuracy eventually led CMC to adopt the corporate slogan, "the Super-Speed People."

"After starting CMC, one of the greatest difficulties we had to overcome involved one piece of technology – the Hydrashift that had been developed at Card & Co., when it was a division of Tuftco. It had come about prior to our leaving Tuftco and was patented technology," Charlie said.

"Not having that when we started CMC made it more difficult for us, especially in the contract carpet segment – making graphics-type patterned carpet. Eventually we were able to create our own Smart Step technology that enabled us to do the same thing as the Hydrashift. We also were able to develop a unique cutting system, needle drive system, the Infinity attachment, Command Performance, and servo mechanisms that we use to control individual strands of yarn."

Computers and small "servo motors" now carry out the functions that clutches and solenoids once performed, feeding the insertion of each strand of yarn across the carpet's width. Shifting needle bars add another dimension to the possibilities for color placement and styling variations. Increasingly, modern-day tufting machines are providing the precision and pattern flexibility that once could be accomplished only on a loom.

Unique, Versatile and Multi-purpose

Every tufting machine built is truly one-of-a-kind, designed to meet the customer's needs and specifications. For that reason, machines that produce the specialized parts for them have been fashioned to provide versatility along with multiple applications. Sonny Jackson, who directs CMC's Jackson gauge parts plant in the tradition established by his late father, James, recalled how the manufacturing process changed because of computerization.

Before computers assumed the role of dictating exactly how a machine should make a part each time, the human mind was the "controller," according to Jackson. He said it required much more skill and training on the part of the craftsman to make parts; quality could fluctuate considerably, largely dependent on the expertise of the craftsman. Today, with the guidance of the computer, each part is fabricated with identical precision.

"We are able to produce tenfold over what we could do when our machines were manually operated – and each part is far more accurate. Before, we would build extremely good fixtures using specialized machines to do what was needed, as opposed to today's machines, which serve multiple purposes," Jackson pointed out.

"The machines now are far more versatile and able to accommodate many different kinds of parts. In a 1950s shop, accuracy was solely dependent on the skill of the craftsman milling the part. If your craftsman happened to arrive at work not feeling well, you knew your parts were not going to be as good that day. Today, it all depends on the accuracy of the machine being used to make it. And machines don't get sick."

Jackson, whose son, James Heath, represents the third generation of his family to devote a career to manufacturing machine parts, said computers have made it possible to perfectly blend precision with speed.

Sonny Jackson (right) and his son, Heath, carry on the mechanical legacy of James Jackson at CMC's Jackson Plant.

"In the early '70s, we built mechanical patterns for slat pattern attachments that were mounted on certain tufting machines. The maximum speed that mills were able to operate them was about 450 rpm. Today, patterned tufting machines operate above 1,000 rpm. The productivity is so much greater and, despite the greater speed, they are far more accurate than the older machines.

"It was like a paradigm shift – when we came up with gauge parts to guarantee accuracy in making carpet, things also changed with the capacity to work at much higher speeds while maintaining the proper balance. And it's amazing to see the capability of our machinery to make gauge parts that are a perfect match so these 'eccentrics' can operate smoothly in tandem, harmoniously."

Even as a longtime insider in the industry, Jackson remains highly impressed by the capabilities of tufting machines today compared to what they were when he started in the industry, "or even what Daddy talked about when I was a kid. We have come from very simple shag or cut pile to intricate designs, patterns, textures and images. And it's amazing just how advanced electronically the industry has become."

Creating a Machine Buyer's 'Menu'

The capacity to produce parts that meet unique requirements, while maintaining extremely high levels of precision, is an obvious marketing asset. When customers contact CMC for new tufting machines, they choose from a "menu" of options, depending on the types of floor covering they plan to produce. Just a sampling of the selections available includes:

- *Organic Pattern Loop Machine* for free-flowing designs, which uses cast modular gauge parts with locator key to ensure uniformity and an in-line needle bar making it unnecessary to step over previously formed back stitches.
- *One-Twelfth Gauge Pattern Loop (1/8 and ¼),* equipped with the Infinity pattern attachment, which creates free-flowing designs with texture and color in any repeat desired. Very popular in broadloom applications, it is the most widely used machine in the carpet tile industry.
- *High Speed Level Cut-Loop* that allows the creation of free-flowing textured patterns with varying amounts of cut and loop, even with a single color, at speeds up to 800 rpm.
- The *Enhanced Graphics Machine* that can create thousands of patterns once available only from a loom, combining precision stitch placement and computerized yarn control with a computerized design center that greatly enhance creativity.

- *Virtual Weave,* which combines loom-like precision with tufting machine speed for a woven look and feel in cut pile or loop pile.
- *Precision Cut/Un-Cut* which uses cut pile and uncut loops to produce thousands of intricate carpet patterns and textures, including geometrics, solids, multi-colors and unlimited combinations.
- The *Infinity* and *Infinity 2E* attachments that offer control of each yarn to produce full-width patterns or repeat patterns of any size desired.
- The *Servo-Scroll* that accommodates up to 144 yarn-feed rolls driven independently to form multiple pile heights at speeds exceeding 1,000 rpm.
- *CMC Quick Thread* attachment that features a rod system with compression rolls allowing for quick, efficient pattern changes.
- The *Yarntronic Pattern Attachment* that can incorporate two and three-pile height capabilities for precise repeatability of styles and patterns.
- *Command Performance,* a computer-controlled servo motor-driven system that automates both set-up and operation of the tufting machine.
- *Smartstep* Shifter, a computer-controlled, servo-driven sliding needle bar attachment that combines instantaneous pattern or style setup with advanced stitch placement and high speed productivity.
- *CMC Advanced Cutting System II,* the most reliable, easy-to-use cutting system in the industry.
- *Gauge Key Modules* to ensure accuracy and repeatability, greatly reducing needle and looper fit-up time; meeting international standards and available in all industry gauges.

Each of these options is light-years beyond anything Lewis and Roy had in mind in the 1950s and early 1960s while they were exploring the frontiers of tufting, first with converted industrial sewing machines and later using early renditions of machines built to accommodate expanding needle bars. In actuality, however, all of today's technological advancements merely represent the latest stage in the logical progression that originated out of the collective genius of the Card brothers and their colleagues.

Major Changes in the Market

Keeping pace with the technological innovations, the market for tufting machines also has changed dramatically. The number of carpet mills in the United States has diminished considerably, from a peak of more than 200 in the late 1960s to the current total of only several dozen. Two giants – Shaw Industries (a significant piece of the Berkshire Hathaway mega-conglomerate) and Mohawk (also with billions of dollars in sales) – account for 63 percent of U.S. sales, according to the May 2009 edition of *Floor Focus* magazine. In Europe, the industry has undergone a similar decline in the total number of carpet mills. Harold North, during his years as CMC's Vice President of Sales, witnessed this change through the years and noted, "There are not a large number of mills anywhere now."

Corporations, rather than privately held businesses, operate most mills today, which has markedly changed the relational dynamics for both sales and service, as North's colleague, Jerry Hendricks, observed.

"In the early days, we formed relationships that were long-term. You were 'in the family' – we were busy and things were always bustling, but it was a much smaller, more closely knit industry. Today, the industry is maturing as it should be and a lot more of the decisions are made at the corporate level rather than by the plant owner, as was the case in the old days. With companies on the stock exchange, you have to take purchasing decisions through a much more complex approval process and evaluation in terms of capital expenditures.

"But when it's all said and done, we still have a congenial, enjoyable business, and maintain good relationships with the carpet mill owners and plant managers. Relationships are still a valuable part of how we do business.

"It has always amazed me to consider how many dollars have changed hands because of sales or deals done with a handshake or over the telephone. There has always been a lot of mutual respect and trust. And for the most part, people have been worthy of that trust and confidence."

Complexity, it seems, is the operative word for every facet of the industry today, according to Hendricks.

"As mills have become more vertical, the focus is not only on tufting machines, but also on all aspects of the finishing process, raw materials and extrusion materials. It's more important to know where every dollar is going.

"For the mills, it's also a more competitive environment. In the old days, if you made a profit, often you really didn't know why. You would try things, throw them against the wall so to speak, and see if they would stick. Today, we understand the market much better and the goal is to produce products that can reach certain price points in order to remain competitive.

"Sustainability is very important, especially since most of the materials (fibers) are derivatives of petroleum products. You have to have all of your i's dotted and t's crossed. You're not only aiming the business in the right direction, but also focusing on the bottom line to survive long-term."

North agreed, conceding that the changes have not necessarily been bad. "Back in the early days, the industry was much smaller. Each of us on our staff would go into the carpet mills and relate on a first-name basis. It was a much friendlier, more informal atmosphere than it is today. The scope of the industry was smaller, not as international as it is today. There was better, much stronger rapport between the tufting machine manufacturers and the mill people, and much stronger, more personal contacts and relationships, all the way to the top with the carpet mill owners and executives.

"I was on a first-name basis with people like Bud Seratean the founder of Coronet; Bob Shaw with Shaw Industries; Ted Munchak, founder of Trend Carpet; Harry and Julian Saul of Queen Carpet Mills, and Shaheen Shaheen of World Carpet Mills. It was mainly business, but we did have social relationships with some people outside of work.

"We were close-knit – the industry was in its infancy, and we were all learning as we went along. Today the industry is much bigger – in terms of the number of people involved and the volume of business, and the tools we use are so much more sophisticated. Computers, word processors, we didn't have them in the early days. I remember doing dictation with a secretary taking shorthand and using old manual typewriters."

Standing by CMC's 300th tufting machine in 1987 are CMC executives (from left) Brian Card, Lewie Card, Danny Henderson, Roy Card and Charlie Monroe.

Curiously, despite the rise of competition and corporate bottom-line orientation, one aspect of the industry that has changed very little is the lack of a protective, proprietary attitude, according to North.

"I learned a long time ago about the value of taking prospective carpet mill owners into existing mills in Dalton. I'm not saying this is the way it will always be done, because next week we could go in and have to do things a different way with new developments and more sophisticated operations.

"But as they used to say years ago, 'There are no secrets in the carpet industry in Dalton, Georgia.' The father of a family might be working in one mill, the mother might be in another mill, and the children might be working in others. Sitting around the dinner table, they might discuss – involuntarily or on purpose – how they did a given procedure.

"The mills have always been open, even to their competitors. They were proud of what they had done and were willing to show it to the world. However, this is much less the case today than what it was. We still have access to some mills, but not all, for taking customers to see one of our machines in operation."

A Small World – Even for Carpet

Although the number of carpet mills has been decreasing overall, reducing the field of prospective tufting machine buyers, new international markets have been opening up since the early 1990s. During the '80s, North's travels outside the United States took him primarily to Europe and Latin America. However, over his 15 years he went to China more than a dozen times, and made several trips to South Africa. Among his other sales destinations for CMC were Australia, New Zealand, Germany, Italy, Spain, Belgium, Luxembourg, France, Denmark and the United Kingdom, and some former Eastern Bloc nations, including Russia, Yugoslavia (now Serbia and Croatia), Czechoslovakia (now Slovakia and the Czech Republic), Hungary and Bulgaria. More Western-oriented nations in the Middle East, such as Kuwait and Dubai, were among the last additions to his travelogue.

Today, international sales have become a major part of CMC's business annually, ranging from 15 percent up to 50 percent or more of its sales. Just as the shift from family-operated to corporate-owned mills changed sales and marketing dynamics in the United States, being able to succeed on the international front has presented its own share of challenges, as Hendricks noted.

"One difference between doing business in America and overseas is that here, you build relationships over a period of years and become acquainted with the people who are able to do the contracts, sign on the dotted line. You deal with central purchasing and the appropriate corporate executives," he said. "On the international scene, since you are not in front of your customers as much, they seem to want everybody in the organization to get acquainted with you. I wasn't always sure whether this was about your honesty and integrity, but there clearly were more people involved in critical decision-making – especially during the initial stages."

A recent trip Hendricks made to Dubai was particularly enlightening. "It was very interesting. Dubai is really growing. They say the country has one-third of all the construction cranes in the world. I believe it – everything over there is under construction."

Adapting to Middle Eastern-Arab cultural mores stretched his comfort zone, he admitted, since business owners in the Middle East were "very tough negotiators. They have grown up bartering and bargaining, and that has become a part of the business culture as well."

Another overall difference, Hendricks said, has involved expectations that most international customers have. "Speed is important to them, but efficiency and versatility are every bit as important. When we sell a machine, we look at production issues and the selling market. Many countries we deal with today, outside the United States, are not as dependent on speed as they are on flexibility. We need to show how our machines can do a specific job they need – at a high rate of speed. Companies in the Third World, for example, want machines that can make maybe as many as 10 different kinds of carpet – 1/10-gauge level loop, 1/10-gauge graphic loop, 1/10-gauge pattern loop, and so on.

"Interestingly, not long ago I visited a company in Australia that was using a machine that had been built in 1973. Our machines were made extremely well even back then, although today's machines are far advanced in terms of speed, efficiency and versatility. But they were still quite happy with our old machine."

During its earlier years, CMC's domestic focus centered around residential machines and contract machines, according to Lewie. But once the base for domestic trade got to the point where CMC felt it should be, its concentration shifted to looking internationally for both residential and contract machines – exploring the export market for both.

"This is driven by the dollar, the Euro and the British pound – international currencies – as well as tax laws. For instance, there was a period when great tax incentives were created in Australia for buying capital equipment, and that stimulated our business over there for that span of time," he said.

Design and manufacturing requirements for international customers are not much different from those in the United States, Lewie said, except in complying with European manufacturing standards – the CE (European

Community) mark. This pertains primarily to the safety aspects of machines, both in terms of materials and how they are configured.

"Styles may be somewhat different overseas, but our machines are custom built to meet our customers' styling and design needs anyway, so that's not a change from what we typically do."

The Challenge of Complexity

Combine the high rate of technological advancement with an ever-transforming U.S. market and increased opportunities on the international scene, and the very exclusive fraternity of tufting machine manufacturers finds itself having to cope with an environment of growing complexity.

A threat of new competition is not the primary issue. Realities of limited sales volume and the specialized expertise that characterize tufting machine makers have restricted participants in the industry to a virtual handful – CMC, Tuftco and Spencer Wright/Cobble (all located in the Chattanooga and northwest Georgia area), along with a British branch of Spencer Wright in Blackburn, England, and smaller, comparatively lower-tech companies in Asia.

The challenge is how to ensure the consistent production and delivery of high-quality machines to customers that have high performance expectations. Although innovation and quality were key areas of emphasis even for pioneers like Joe Cobble, and Lewis and Roy Card, they worked in comparatively simpler times, unencumbered by precedent. The machines they made were much less complicated. With fewer employees to deal with, personnel issues were not nearly as vexing. Customer service was always a concern, but with fewer variables. And worries about the global economy were non-existent – at least early on.

Once CMC began to build momentum and formally made its entry into machine manufacturing, principals Charlie and Lewie recognized the need to establish well-considered, clearly articulated guidelines for everyday business operations. Just as a ship uses an anchor to hold its position in a harbor, over the years CMC also found that corporate "anchors" would help to hold it steady in the face of the often shifting, unpredictable winds of industrial, fiscal and global change.

Becoming the family's third generation in the industry, they also wanted to ensure that they would always build on the solid foundation that had been laid by the Cobbles and the Cards. One element of this has been continuing to involve family members in key leadership and decision-making roles at CMC.

Lewis and Roy Card stand in front of a wall at CMC that
displays plaques for some of their many tufting patents.

As noted earlier, Lewis and Roy continue to serve with Charlie and
Lewie on the board of directors, offering their wisdom and perspectives from
watching the industry evolve from its inception. Roy's son, Brian, CMC's
CFO and treasurer, and Lewis's daughter, Janice Card Henderson, also
participate as members of the board.

Broad Involvement in Leadership

However, building on their rich and storied heritage, while maintaining a
leading role in the industry, has demanded much more than keeping control
of the company within the family. Operational roles were clearly defined.
Charlie, now Chairman and President of CMC, is heavily involved in the
customer side of the business – sales and marketing – as well as the Creative
Tufting Center. He stays very active in service and product, the external
side of the business. Lewie, CMC's Vice Chairman and Secretary, is in
charge of operations and finance, the people side and the internal side of the
company.

They have taken great pains, however, not to impose leadership from the
proverbial ivory tower. Charlie and Lewie are part of a 14-person management
team. Other current members of the team are family members Ryan Berube,
Brad and Brian Card, Keith Askew and Zach Monroe; plant manager Allen
Neely; Sonny Jackson, head of the Jackson machine shop; Rick Howard, the
materials manager; Jim Joyner, director of lean, quality & service; David
Lynch, the engineering manager; and Jack Hatcher, consultant in strategic
planning and human resources.

Continuity has been a major plus, Charlie noted. "Since the first year we started functioning as a management team with our initial strategic plan, we have not lost anyone from our original management group, but have added people as the business has grown and developed.

"Each member of the team has a strategic function in the company – they are department heads. We meet as a group a minimum of twice each month, both as a leadership team and strategic planning team."

Candor and the free flow of input during each monthly meeting are strongly encouraged. "Over the years, everybody has gotten more and more comfortable with voicing their ideas and concerns," he said. "When we first started (meeting as a team), things were pretty tense and tight. But our relationships have grown as we have come to trust each other more and more over time. Everyone has seen that we truly value what the other person brings to the table."

Lewie added that CMC's emphasis on individuals and relationships "has enabled us to be a healthy, functional family rather than an unhealthy, dysfunctional family. They know that our approach to business comes from the heart and that we are people who truly care, not people who don't care about them."

"We are thankful for the men and women of character that God has assembled here. It makes it so enjoyable coming to work every day. That doesn't mean we are all alike, not in any sense. We are all unique individuals, and God has given us the primary responsibility – and privilege – of leading the people here," Charlie said.

Articulating the Corporate Mission

Beyond that, in recent years the corporate leadership has collaborated to articulate in writing its goals, standards, values and beliefs. Presented in the form of the CMC Mission Statement, Values Statement, and Quality Policy, these documents are regularly reviewed by company leaders and staff, and are also available to customers and suppliers. They serve both as communications instruments and tools for accountability throughout the organization.

"We had been consulting with Jim Joyner, an authority in the concept of Total Quality," Charlie said. "He started helping us in 1993 with the process of hammering out a formal quality policy for the company. It seemed like it took forever, but we finally arrived at a statement we all were satisfied with.

"About two or three years later, in one of our strategic planning sessions we decided that we needed to draft a mission statement. We worked really hard at that, wordsmithing it, and eventually as a team we came up with something we felt good about. In 1996 what we arrived at was this:

> *Our mission is to be the tufting equipment*
> *supplier of choice throughout the world,*
> *by developing and nurturing partnerships with customers, employees,*
> *and suppliers while achieving an adequate return.*
> *– All to the Glory of God –*

Over the succeeding months, within CMC each department also went through a process of drafting its own specifically worded mission statement to complement the broader corporate declaration. For example, the Creative Tufting Center has kept the following statement clearly posted as a daily reminder to its staff:

Our mission is to support the sale of tufting machines by developing new sewing applications and tufting technology; producing samples that tell the marketing story for new products; partnering with customers in the design and development of their products through education and demonstration of machine capabilities; and providing post-sale support by assisting the CMC Sales and Service Departments with specific customer needs.

"Today each department has its own mission statement, which works toward fulfillment of our corporate Mission Statement," Charlie pointed out.

"Governments provide minimum standards for all businesses, but we understand that God holds us to an even higher standard. We wanted to have in writing the values and formalized statements to express what our ideals are. This helps to provide an objective way for evaluating what is going on at CMC.

"Our Mission Statement says everything we do is 'All to the Glory of God.' That is ultimately who we answer to. Because He holds us to a higher standard than the laws of the land, or even established ethical practices and mores of society, we hope that our customers and our employees will receive the benefits of that.

"Even in writing, it's a very difficult thing to communicate throughout the business. We want it visible everywhere in the company as a constant reminder. That's why it's posted throughout the building, along with each department's own mission statement."

CMC Puts Its Values into Writing

Many companies have established mission statements that express their intended purpose or overriding goals. However, in addition to its own Mission Statement, CMC's management team in 2001 initiated steps to adopt what it called its Values Statement, identifying the intrinsic values that corporately – and individually – the company, its leaders and staff would strive to embrace and demonstrate in a consistent manner.

"About five years after we adopted our Mission Statement, we determined that it would help if we clearly defined our values. These first were formulated by Lewie, Brad Card and myself, and then reviewed and acted upon by our leadership team," Charlie stated.

The following is that Values Statement:

WHAT WE VALUE MOST

As we relate with our customers, suppliers and each other:

Honesty and Integrity
We take great care in making commitments, and then we live by them. We require honesty and integrity in everything we do.

Love, Dignity and Respect
We look for the best in others and see each person as a unique individual. In all our dealings, we will treat each other with courtesy, respect, consideration, and acceptance.

Servant's Heart
We will serve others by:
- *Having a cooperative spirit*
- *Having no hidden agendas*
- *Having a humble disposition*
- *Thinking of others first*

If you don't believe that we are living up to these values, please let us know. If you think we are, we would like to hear that also.

"In each case, there are times when we fail miserably to live up to what we have stated, more times than we would like to admit, but that is where our heart is and we are not satisfied unless we are living it out as best we can," Charlie observed.

"I have seen Lewie, time after time, if a meeting is going on and a conflict arises, stop and read our Values Statement to remind us of what we're there for. It's amazing how that gets us back on track every time."

The Values Statement has also served as a tangible resource for employees to refer to in understanding how they are viewed – and valued – by their superiors.

"Once we adopted this statement, our staff began to see that we intend to operate just as our values statement says – with a servant's heart, having a cooperative spirit and thinking of others first – and that we don't have any hidden agendas," Charlie said.

For many of the employees, this was a welcomed change from some of their experiences with previous employers. "It primarily goes back to our values statement – we value what people think and what they have to say; we know they each are unique individuals who have earned the right to be heard, and we seek to treat each other with love, dignity and respect," Lewie commented.

"I often think about Joe and Albert Cobble. Those two brothers couldn't even be partners for two years. Charlie and I have worked together for 28 years, and Lewis and Roy have done it for nearly 60 years together. To succeed, it's like a marriage – it requires a lot of give and take.

"That also applies to the people who have been with us here at CMC for 20, 25 years or longer. We spend a lot of time together and have to interact with one another a great deal. Qualities like personal humility, having a cooperative spirit and thinking of others first make a tremendous difference.

"I have been approached a number of times over the years and questioned about the way I handled a situation or dealt with a person. It's subjective, rather than objective, but it keeps us aware of how other people see us. We definitely don't want to go by a common distortion of the Golden Rule – the idea that the one with the gold rules."

The closing sentence in the Values Statement is one that typically catches first-time readers by surprise, according to Charlie. "The most important part of the Values Statement is the last part: 'If you don't believe that we are living up to these values, please let us know. If you think we are, we would like to hear that also.' Lewie and I still have feet of clay, and our motives are not always pure. But that is why we have our values expressed in writing – we want to know about it if we're not living up to them."

Quality – a Commitment, Not Just a Concept

Clarity of mission and succinct articulation of values have helped to communicate CMC's standards and corporate expectations within and outside of the company, but a third key area needed to be consistently addressed as well: *Quality*. It is one thing to believe in quality and strive for it, but achieving and sustaining it at high levels presents a different challenge altogether.

"It's important to remember that in terms of the development of tufting machines and the heritage that CMC has built on, my dad and his brother Roy did everything in a quality way," Lewie noted. "The difference was they did it in a more hands-on, innovation-by-innovation, machine-by-machine manner. They did not establish a formal quality initiative, but it was not necessary then. Today, due to the complexity of the machines we produce, all the continuing technological advancements, and the ever-expanding international scope of our business, establishing a formal quality process has become much more critical.

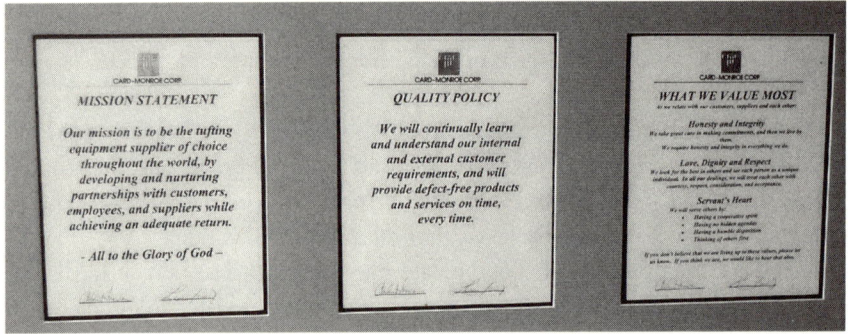

The CMC Mission Statement, Quality Policy and Values
Statement all are prominently displayed throughout the
company's plants for customers, employees and suppliers.

"Today the greatest advances in our industry are taking place through human/machine interface, discovering how different machines can be used to make new things, resulting in outcomes, controls and consistencies that were never possible before. And all of this is without even improving the mechanisms – which we are always striving to do as well. So taking intentional, formal steps to ensure quality has been more crucial than ever.

"I don't want anyone to think that quality was an idea we came up with. There has not been a man or woman who passed through the doors of one of the tufting machinery companies that hasn't made a difference in the industry. Every person who has been involved deserves a share of the credit for what has come about in the tufting industry. They are all part of the legacy that we are continuing to build on today," Lewie said.

"Granted, we have more sophisticated tools available to us in the 21st century, but all we are doing is taking principles and techniques that were established decades ago and putting them to proper use through the advances that have been made through the years.

"As I said, the focus on ours being the tufting equipment of choice is a result of the legacy we inherited from Lewis, Roy and others, and our desire is to remain leaders in this industry. All we are doing is using modern tools to do what they taught us to do."

Exercise in Defining Terms

One challenge CMC faced in drafting a formalized expression similar to its mission and values statements centered on defining exactly what the company meant when the term "quality" was used. Like beauty, what quality actually looks like tends to lie in the eyes of the beholder. The leadership team, however, wanted something more concrete and specific.

To this end CMC began its consulting relationship with Jim Joyner in 1993. He brought extensive training and experience in the area of Total Quality, which developed out of a variety of industrial quality movements in the 1980s and early 1990s.

After 11 years in the 3-M Corporation's quality program in St. Paul, Minnesota, Joyner had teamed with quality guru Phil Crosby at The Quality College in Florida to train executives and managers in quality principles and managing quality, as well as consult with a number of large corporations. In 1991, Joyner started his own consulting company and found himself in demand, working with companies in many parts of the world.

In 1993, wearied by the rigors of international travel and wanting to spend more time with his family, he elected to pursue business closer to his home in Chattanooga. Through a mutual friend, he was introduced to Charlie.

"Until then I had never heard of CMC, but I called Charlie and he said he would be glad to meet with me. As we met, he stated they had been looking for help in the area of quality. They had been to seminars and seen tapes on quality, but still were not where they wanted to be. Our 15-minute meeting ended with a handshake – and we have been working together ever since."

One of Joyner's first steps with CMC was to take the leadership team on a weekend retreat at Dunaway, near Dunlap, Tennessee, where Lewie had a cabin. "We spent the entire weekend learning about Total Quality," he recalled.

Soon afterward Joyner established a "train the trainers" program at CMC, equipping the management team in how to take Total Quality principles they learned and apply them to the business, as well as how to teach the concepts to other staff in a practical, structured way.

CMC remained one of his consulting clients until 2004, when he was hired full-time to become the company's first director of lean, quality & service.

But Joyner's value for CMC was established long before he accepted a permanent staff role. With him serving as facilitator, CMC's leadership team met a number of times to hammer out a Quality Policy statement. "It seemed like it took forever," Charlie said, "but we finally arrived at a statement we were happy with." It reads:

"We will continually learn and understand our internal and external customer requirements, and will provide defect-free products and services on time, every time."

In particular, CMC desired to place special emphasis on phrases like "defect-free" and "on time, every time," making them more than lofty ideals – they wanted them to become everyday realities. It was agreed that merely seeking to minimize defects would not be acceptable; prompt delivery, as promised, would become a hallmark for the company.

To transform the Quality Policy from rhetoric to reality necessitated a strategy, and Joyner has served in the role of developing and helping to implement the strategy on a company-wide basis.

Centering on Cycle Times

"A fundamental question we dealt with was how long it took from the time we would get an order for a machine to the time that it was ready to be physically shipped – what we defined as the 'cycle time'," Joyner said. "This included custom designing the machine to the customer's specifications, making the parts, building the machine, testing it, and finally having it transported to the customer. Customers care about and feel our lead times, so shortening them affects customer satisfaction.

"We also asked, is the amount of time required different for a relatively simple Type 1 machine compared to a more complex Type 5 machine? We learned that understanding the cycle times – following the manufacture of specific machines from start to finish – was a key part of understanding our business as a whole.

"As we tracked this, we began to realize that in every instance when a cycle time is extended, that means problems along the way have affected the schedule. So we began to analyze segments of the cycle time to see if anything in the process was inherently wasteful. This was where we began to apply the concept of 'lean manufacturing' – a method to eliminate waste and non-value-added activities in the organization," Joyner said.

"Studies have shown that in most organizations, only 10 percent of all activities can be considered 'value-added.' The other 90 percent of activities are there largely because management doesn't know any better. You will never completely eliminate waste, but we reasoned that even if we could raise our ratio from 10/90 to 20/80, that still would be twice as good.

"One step we took toward achieving that was the WOW – Waste Observation Worksheet. Every employee was asked to become more sensitive to things going on around them that were wasteful and could be improved."

Jim Joyner (left), CMC's Director of Quality, admires a carpet
sample with Brad Card, Director of Marketing.

Since adopting Lean Manufacturing in 2003, CMC has been implementing numerous changes – from the time sales orders are taken, through the manufacturing of parts and assembly, to the time when tufting machines are fully assembled and shipped. Improvements throughout the company have been dramatic, and customers have responded positively and enthusiastically. Comments they often express include phrases such as "fast to respond," "integrity," and "trustworthy – they do what they say they'll do."

Not that CMC believes that in any sense it has "arrived" – there is always room for further improvement. "CMC is hitting on all cylinders right now, but we have to keep fine-tuned at all times." Lewie said. "They say the cream always rises to the top, but we don't want to get caught reading our press clippings – we don't want to take our success for granted or become complacent.

"Daily, even minute by minute, we never want to take lightly our commitment for doing things right – and you can always do things better, which we are striving to do."

And the Quality Policy remains a clear and visible guide for objective evaluation.

"Every year in our strategic planning meetings we review our Quality Policy to see if it still works. It has not changed since it was adopted," Charlie said.

Even with this policy, however, an overriding philosophy serves as paramount motivation throughout the CMC organization. It is a commitment, starting from the top on down, to "doing things right." The Quality Policy is a centerpiece of this philosophy, but there is much more to it than that, as we will see through the eyes of a number of key individuals.

These busts of Roy T. Card, Joe Cobble and Lewis Card, Sr. greet guests in the lobby at Card-Monroe Corp. The sculptures were commissioned by wives of the company's principals to pay tribute to these true pioneers of the tufting industry.

12

DOING THINGS RIGHT

Imagination and innovation are crucial to the success of a cutting-edge company like CMC. But "new and exciting" can take a company only so far. You still have to deliver the goods. That is why their overriding concern has focused on how to ensure high-level performance for their customers. This has included building specialized machines capable of producing an ever-changing and expanding variety of residential and commercial carpet with the speed and precision expected, providing quality service whenever needed, and conducting business in a fair and equitable manner.

No company operates with the intention of sending out substandard products. Consistent quality is essential to retaining customers and increasing market share. For this reason, many companies have eagerly embraced and utilized concepts such as quality control (strategies for minimizing defects) and so-called "quality circles" (including employees at all levels of the manufacturing process in planning and policy-making).

At CMC, however, this quest for quality has led to a significantly different approach. Emphasis on "Total Quality" has shaped a comprehensive philosophy and strategy involving every aspect of the company and its staff. Achieving this goal essentially has amounted to a merging of the company's Mission Statement, Values Statement and Quality Policy into a framework for fulfilling a simple commitment: Doing things right.

Doing What They Say

A 2007 third-party survey of CMC customers elicited a number of comments indicating that the company's resolve has become more than just rhetoric. These included evaluations such as, "they stand behind their products," "they do what they say," "CMC's service is the best of the best," and "they work hard, always trying to do the right thing."

CMC's leaders find such glowing reviews heartening, but they fully understand and appreciate the hard work and determination required to merit such responses.

"There are no givens in this business." Lewie Card observed. "Day in, day out, we are committed to doing the right thing, taking care of our suppliers, customers and employees. We want to do the right thing, building and strengthening relationships, no matter who we are dealing with. We want to do what's best for them, even when it's not always best for us. It's a way of applying the so-called Golden Rule: Treating others the way we would want to be treated if we were in their position.

"You have to stick with the basics: quality, service and price. We know we are not going to be the lowest in price, because we feel we have to put a lot into our machines – extras that go above and beyond what is necessary. So quality and service will always be what sets us apart – our distinguishing features," Lewie said.

"For example, we use the highest grade automotive paint that you can buy for our machines. Lesser paint might initially look as good when a machine is going out, but it won't look as good down the road. To express it in colloquial terms, we don't put mud flaps on our Cadillacs. We make sure to do it right the first time. Putting in extra quality has a cost, obviously, but it has paybacks in how long our machines last and that operational uptime is maintained at an absolute maximum.

"There also is a very large expense in machine development that people don't see, and we have to get a return for that. It's not just about building the product, but also getting to the point where we can manufacture in a way that guarantees the quality of the product."

Not Just Quality, But 'Total Quality'

This commitment has served as the genesis for CMC's "Quality Policy," as introduced in the preceding chapter:

"We will continually learn and understand our internal and external customer requirements, and will provide defect-free products and services on time, every time."

Lewie, Charlie Monroe and CMC's leadership team have discovered that the "how-to" behind this policy can be captured in two words: *Total Quality.* A simple phrase, yes. But for CMC it represents a scope and breadth of meaning that encompasses the entire company.

Even in the early days, from the time Lewis Card began working for his uncle, Joe Cobble, and later when Lewis's brother, Roy, joined the business and began contributing his skills, quality and excellence were goals they continually strived to attain. But as technology gave birth to faster, more complicated machines, quality issues also grew in complexity. Nevertheless, CMC resolved to make concepts like "defect-free" and "on time, every time" more than catch phrases – they wanted these to reflect the company's everyday operations.

A pivotal issue for CMC has involved assessing how well the company was performing in its various departments. "It was very difficult to get an accurate assessment of performance from managers or employees," Lewie said. "I typically would ask, 'How do you all feel that we're doing,' but answering a question like that is very subjective."

Finding ways for measurable, objective evaluations was one of Jim Joyner's first challenges upon arrival at Card-Monroe as a consultant.

"We had grown so fast, adding on to our plant seven times in five years, we were afraid of going out of control if we didn't take proper steps to prevent that. We wanted to ensure always being able to provide the best for our customers," Lewie recalled. "Jim helped us to move quality from just an emphasis among our leadership team to a full operational initiative. It was not limited to the production areas; it also started at the receptionist's desk, our administrative assistants, clerical staff and so on. It meant awareness that quality is involved in everything you do. But being aware of it, thinking about it, is one thing; living and breathing it is another story."

"We wanted our business to be all that it could be, and realized that with a focus on quality it could be so much better – not only in terms of our products, but also in the processes that we used throughout the company," added Charlie.

In that light, Joyner explained, "My job was to find ways of converting feelings into facts, using measurement criteria we all could agree on."

Among the strategies he introduced were the concepts of cellular and lean manufacturing.

Trusting the Concepts, Committing to Them

"A lot of this is counter-intuitive," he said. "You have to be convicted by the concepts, catch on and trust them. It's not how hard I am working, how efficiently I am doing it or how fast we are moving. The issue is cycle time – how long it takes from the time we receive an order to the time we ship the machine, the customer receives it and gets it installed and operating."

The thinking behind cellular manufacturing can run counter to long-revered practices and paradigms, but Joyner noted it offers numerous benefits leading to increased productivity and efficiency, including:

- Elimination of "dead time" (fewer hand-offs)
- Less motion (machine and crew moves are less)
- Making work easier (less overhead work)
- Making work faster (better assembly sequence)
- Enhancing ownership (expanded responsibility)
- Improving communication (better teamwork)

This concept has had a tremendous impact at CMC, according to Joyner, reducing the overall cycle time for manufacturing a tufting machine – from start to finish – by approximately 50 percent "That is a huge chunk of time," he pointed out.

As a result, CMC has become more responsive to customers and more "agile" than ever before. "It's like wellness and fitness," Joyner stated. "When you're fat and slothful, you can't move as easily. As a company, we are leaner and fitter than we have ever been – and we're not done yet. There are some significant things we are involved in now in the production area that will have a great impact on our ability to serve our customers even better.

"The purpose, the objective of lean, is to eliminate waste and add value to the customer. So cycle time becomes your trump measurement: If you see your cycle time extending, it automatically tells you that you are encountering some type of waste. You may not know what it is, but you realize that in some way, waste has just increased."

He has presented these ideas to the Chattanooga Manufacturers Association, and in 2007 CMC was named one of the 50 recipients of the Progressive Manufacturers Award, bestowed annually by *Managing Automation* magazine, published by the Thomas Register, producer of a comprehensive listing of manufacturers for many different products.

CMC has seen measurable results in many ways, including:

- On-time delivery of new machines has doubled.
- Manufacturing cycle time has been reduced by 50%.
- Productivity has increased by 45%.
- WIP (Work in Progress) inventory has been reduced by 70%.
- Nonconformance has been reduced by 20%.
- Service parts shipped on-time (by request) has risen from 66% to 85%.
- Labor hours, comparing 2005 to 2004 for example, have been reduced by 15%.

Profitable – and the Right Thing to Do

The motivation behind CMC's emphasis on "lean" is not just profitability, Joyner asserted. "This is all being done at CMC because it's the right thing to do. We believe we are stewards over the resources given to us. In response to the business we are honored with by our customers, we owe them the best possible machines we can give them. In our corporate culture, waste is undesirable and inconsistent with the values of our organization. Total Quality is important not because we *have* to do it, but because we *want* to do it. And again, because it's the right thing to do.

"People like working in an organization that is operated well, and our customers can see it and feel it," he said. "We received a survey response recently from a new European customer who had finally ordered a machine from CMC. 'This is a Rolls Royce,' he wrote. 'I have been driving junk!' He said he is now one of our biggest proponents. The only thing he wonders is, 'Why didn't I do this sooner?'

"I have learned so much since I have been here at CMC," Joyner said. "I had known about some of it, but I really did not understand it until I started working at it and applying it in a day-to-day environment. Now I know that if you're not doing lean, you're not doing Total Quality."

Clearly Joyner was not the only "student" in this process. Everyone at CMC has been learning, from the top on down through the ranks.

CMC staff members stand behind a banner affirming the company's commitment to Quality in 1998. (Photo by Med Dement)

"An important portion of our quality initiative today is lean," Charlie stated. "But contrary to what one might commonly think, this does not mean reducing the number of people – it means reducing waste. This can be in terms of excessive movement, doing unnecessary steps, or having too much or even the wrong inventory.

"From that standpoint our lean initiative has helped us to become more agile, better able to meet our customers' expectations. Price is always a consideration, obviously, but our number one concern is service and quality. We have a saying around here that we only want to take care of the customers we want to keep – and we want to keep them all!"

One of the key concepts stressed in quality training, according to Charlie, is the principle of *suboptimization*. "The tendency is to create an imaginary wall between each department. If we don't encourage open dialogue, we typically make decisions based on what we know and what makes sense to us – what's best for us – regardless of whether that is actually what is best for everyone involved. This is called 'suboptimization' and presents great danger.

"Any decision that optimizes one process to the detriment of another process results in suboptimization. Once we understood this, we could see immediate benefits – but it's not a one-time exercise. Avoiding suboptimization requires a different mentality, a new way of thinking about what we do in relation to others. Almost overnight it began to change the culture of our company, helping us to become more professional and positive in terms of how individuals and their departments interacted."

"The end result of this was keeping our focus on our ultimate customer and making sure that their needs were met," Lewie noted. "In our case, the ultimate customer is the carpet mill, not the person who buys and walks on the carpet. The carpet mills do their job, learning from their customers – consumers – and as we learn from the mills, we can do a better job of helping them."

"The overriding theme of Total Quality is continual improvement," Charlie said. "What we are doing in the areas of lean is part of that ongoing improvement. If we are continually improving in one phase of the business, it bleeds over into other areas.

"Is everybody at CMC totally aware of this emphasis? No. Even I at times may be guilty of suboptimization. It requires steady plodding, along with occasional jolts along the way to wake us up when we have strayed off course."

Quality Cannot Be an Add-on

"If you're not making quality a central part of the process, there is no way of back-engineering quality into the product," Charlie asserted. "The process guarantees the product. It's getting 10 good products out of 10 every time, not attempting to minimize substandard products by inspecting after the fact."

Lewie agreed: "You can always inspect and correct, but that's very costly – there is a lot of waste, in terms of both materials and time. So for us the key is making quality integral to what we do, from the time we make the initial call on a customer to sewing off the machine after delivery and installation – and everything in between."

"Quality is a forever process that you never perfect, but you keep working at it. Like anything you want to improve, you don't just do it once and presume you have it down – that applies to playing tennis or golf, building a marriage, being a father, or running a business. You work every day to get better and better at what you do."

Consultant Jack Hatcher, having worked with CMC since 1994, attested that the company's pursuit – and achievement – of quality has not come about by chance.

"By no means am I an expert in the carpet industry," he said, although his clients have included Mohawk Carpets, Textile Rubber & Chemical, which does business with the carpet industry, and the Carpet & Rug Institute. "I work with companies in the planning process, and CMC is definitely a strategically planned organization.

"CMC takes a thorough look at itself annually – strengths and weaknesses, opportunities and threats. In planning sessions, CMC's management team prioritizes the goals and strategies to be accomplished during the next fiscal year. Their planning culminates with due dates and names, reducing it all to print, to be communicated to the employees by the CEO in presentation form annually. Then key managers meet monthly to report on progress being made on assignments drawn from the strategic plan."

Product, Quality and Customers First

"There is no question in my mind that the management of this company places product, quality and customer satisfaction above profit," Hatcher stated. "I have served on a number of corporate boards of directors, and consulted with many more companies, and can assure you that profit being placed in a secondary position is highly unusual."

He has served on the boards of two privately held companies that are both extremely growth-oriented. "There is no question that with these companies, sales volume and profit goals are drop-dead serious. Their evaluations of success or failure every year are predicated on whether they have met their sales and profit goals.

"At CMC, by contrast, success is constituted by interpreting customer expectations, offering fresh innovations to them, shipping the order on time, and sewing off the machine to operate perfectly. All the while, CMC also is entertaining the machine owner's product designers at the Creative Tufting Center, assisting them in understanding the capabilities of their machines so they can incorporate those designs."

Hatcher himself has felt like a primary beneficiary of CMC's people and customer-oriented focus.

"Although I am not a full-time employee, I really feel like a member of this (corporate) family. I would describe the organization as 'paternal,' even though that is not a good term the way it is commonly understood in the business world.

"CMC is a truly caring organization, and the actions and attitudes of the management are drawn from their commitment to complying with their corporate statements of mission, vision and quality that have been expressed in printed form and posted on many of the walls in this building to serve as a constant reminder."

Building around Relationships

As might be expected, members of the CMC management team have unique perspectives on the company's determination to do things right, influenced by their responsibilities and the areas where they provide leadership. For example, Keith Askew saw it manifested at the Creative Tufting Center that he manages, serving as a liaison between technicians at the Center and customers.

"We try to nurture our relationship with the designers, making sure they are getting what they want from our machines. If we can help make them successful, that makes them heroes when they return to their companies."

Askew said CMC articulates very well its philosophy of doing things right. "I think of it in terms of machines, salesmen, technicians and my department. I'm very aware of our company's image and commitment to quality, and we try to satisfy customers in their quest for developing new products.

"We want them to feel comfortable, knowing they can trust us to perform at our best when they are here. A strong relationship with them helps us in meeting their needs better and at the same time enhances our chances of retaining their business. We want them to see how easy it is for them to achieve the type of products they want to make.

"It starts with the designer," Askew said. "We show them how to apply their design to our machines to get the carpet, the look that they want. Our operating system is easy to use, very designer-friendly. We show them how simple it is to take their pattern and make carpet with it. This helps to take away any fear or headaches they may have about trying to make a new product.

"Because the designer who comes here gets so much help mechanically and technically, we have some customers that return frequently to design their product lines. They can focus on being creative, working at their own pace, without being limited by time constraints they often encounter in a plant production environment.

"In describing their relationship with Card-Monroe, 'support' is a word I hear customers use a lot. They feel like we really support them when they have an issue, whether it involves the machine or something they are trying to get the machine to produce. CMC – my department, the service department, sales – we are all willing to do what is necessary to satisfy and fulfill their needs to the degree that it's possible.

"For us, it's a source of pride – in our own workmanship, as well as being able to maximize the capabilities of our machines. We feel a sense of ownership in doing the best that we can do. In the past, Lewis Sr. and Roy set the pace, providing the example for hard work and doing their best, and now Charlie and Lewie are carrying that on."

Simple Goals Despite Complexity

David Lynch came to CMC to become its engineering manager in 1996, after 15 years with Duracell Batteries. His primary responsibility has been to take customer specifications from the sales team and turn those into documents that CMC's staff uses to make parts and assemble machines, based on the customers' requirements.

Doing things right in building a new tufting machine requires more than following a basic template, he noted. "We have very few opportunities where the customer calls back and says, 'I want a machine exactly like the one you sent me the last time.' They usually say, 'I like the last machine, but…,' and they tell us what needs to be different on the next one. These differences may be small to major. Although there may be some similarities, the vast majority of machines we make are unique in some way."

Over the years CMC has incorporated hundreds of variations into the machines, beginning with six different sizes of machines just based on sewing width – ranging from the 42-inch sample machine up to 204-inch production machines. There are many advantages to offering virtually limitless variations, according to Lynch, such as great flexibility in meeting customer needs. At the same time, however, this increases the challenge of doing things right and avoiding problems.

To put this complexity in numerical terms, the "simpler" machines that CMC produces – turf or other high-pile machines – have 650-700 different parts and a total of more than 5,000 individual pieces. More complex machines, with the latest in pattern attachments and computer-control technologies, have 850-900 different parts and total more than 34,000 separate pieces, including large quantities of needles, knives, loopers, servo-motors and rolls.

"Our cellular approach in engineering and the idea of lean manufacturing have caused people to think more about our process flow and ways that it can be improved. It has made for a smoother process, and our output is much better matched to the capacity of our downstream 'customer' – moving from production control, where jobs are started, to the machine shop, to the assembly department," he said.

Above is a view of CMC's final assembly area using traditional production methods, while below is the cellular assembly area, where product is built from start to finish by a single team.

"It's no longer a matter of just getting your own work done and sending it down the line. It does not help to say, 'We got our part done,' if the people in the next step of the process are not ready for it. For instance, in the machine shop it makes no sense to run six months worth of parts if manufacturing is not ready to use them. Why tie up a machine making parts that will be needed only at some point in the future when you can be working on a part that is needed now?

"We have been learning to rethink the whole idea of efficiency, looking beyond the borders of our own departments. It runs counter to traditional thinking, but overall the company is more efficient if one department is a little less 'efficient' but is working in harmony with the other departments."

The idea of coordinating productivity from one department to the next "makes you feel just a little more closely connected to the folks downstream," Lynch said. "When your work is done, you know someone is ready for it and prepared to put it to use in the next stage of building the machine. We have a real feeling of doing our work for him or her – the next person in the process – rather than an attitude of, 'I'm just doing this to get it off my plate,' dumping out a bunch of work that will just go into inventory somewhere."

Poised to Correct Any Wrongs

Brad Card, now Director of Marketing for CMC, has been with the company for more than 20 years, since February 1989. He recalled a situation illustrative of doing things right with one customer, Aladdin Mills/Mohawk in Dalton. "I had quoted a machine to them and as I was writing up the quote, noted that we had sold several similar machines to them. I realized that on the last one they bought from us, we had inadvertently overcharged them by about $5,000. I went to tell Charlie about it, explaining the machine was already in place and had been paid for. He immediately said we should call the company's representative, telling them we had made the mistake and would credit them the amount.

"The representative, Joe Yarborough, was appreciative and responded that was one of the reasons he did business with us, knowing that he could count on us. He assured me that he had implicit trust in us, that if we made a mistake we would make it right.

"That is the kind of environment that we work in here," Card said. "My role is to do the best job I can in taking care of our customers, giving them what they need to succeed in making carpet. We feel we have the best product, a solid product known around the world. Once we get a machine in the plant, we do all we can to help the customers – especially the hands-on customers, the ones that will operating or maintaining the machine – to be successful.

"Our machines may sell themselves, but we also need to provide prompt service. When a customer calls and asks for help, we need to respond quickly, abiding by the philosophy, 'The customer is always right.'"

Doing things right, Brad Card said, extends beyond the company culture. It also means conforming to customer requirements, their definitions of quality. "We have learned that we can do a lot of things right, doing all we can to build a machine the right way, but if it doesn't meet the customer requirements, nothing else matters.

"It starts with our quality initiative, our commitment to do a better job. But then we have to stop and ask, 'What does that mean?' These all play a part in making the company what it is. We are constantly looking at our processes, seeking to improve internally. This includes looking at numbers, at actual data. Are we improving in our ability to service our customers and succeeding in satisfying their needs?

"Charlie's goal is simple: 'to create positive customer experiences.' And we think one of our greatest opportunities for doing this is when customer representatives are in our 'home' – our plant. We strive to make them feel welcome, with the desire to make them look good in doing their jobs. If we can help them be successful, then we're successful."

Seeing Proof in Action

Ryan Berube has been with CMC since 2000, his first job after graduating from the University of Illinois in electrical engineering. He started in the research and development department, then proceeded into the engineering department. As assistant plant manager, he is responsible for oversight of the machine shop, production, maintenance and electrical departments, and is involved with the research department.

Regarding doing things right, Berube said, "My goal is to get a defect-free product out on time. With me, this is the finished product – the tufting machine itself. It's challenging, but we strive to achieve this on every machine."

"We have a lot of continuous improvement initiatives to reduce non-conformances. These can include design flaws, parts that were made incorrectly, something that arrives late and delays production, or if the assembly crew does something wrong while assembling a machine.

"My strategic goal is to devise any way possible to reduce the number of non-conformances, whatever they may be. These affect our cycle time and shipping dates, our ability to keep our promise on delivery."

Established practices and old paradigms do not die easily, however, Berube noted.

"Through our emphasis on Lean Manufacturing, we have learned a lot. Admittedly, our initial reaction when the concept was introduced several years ago was extreme skepticism. We felt it was just another fad program being implemented for a while that the company would eventually abandon.

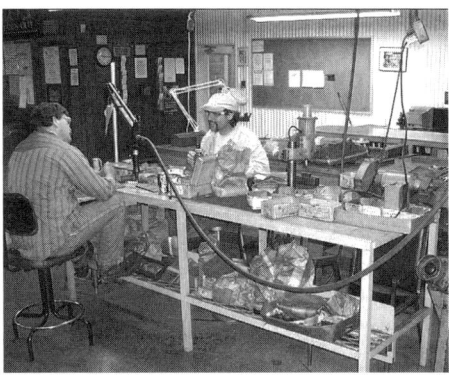

The top photo shows the gauge part assembly area at CMC's Jackson Plant before a formal lean and quality initiative was instituted. Below is the same area with a well-defined, carefully laid out subassembly area to enhance worker efficiency.

"But the results have been tangible, and now our guys are much more receptive to change. It involves all of our processes, even the physical layout of the plant," he said.

"Today, our employees are much more excited about what's going on. Four or five years ago, they were very doubtful that it would work. It has had a tremendously positive impact on our production efficiencies and cost effectiveness.

"The idea of lean manufacturing was all new to me, coming from an engineering background. Typically when we hear the word 'lean' we think of cutting back on staff and overall expenditures. But in fact, since we began implementing this, we have grown in our head count to keep pace with the increased production demand. Lean isn't about cutting back; it's strictly about eliminating waste in any form."

In terms of overall CMC core values, Berube said the commitment to doing things right has resulted in "placing the emphasis with our employees, allowing them to be more productive and become more satisfied in the day-to-day tasks that they perform. We want their work to be more fulfilling and easier, and our emphasis on lean manufacturing has enabled us to focus more on that.

"Another element of doing things right is that I want to put the best people in the position to maximize their talents for the benefit of Card-Monroe and its customers. When we do that, it results in satisfied employees, shareholders and, most importantly, our customers."

Bringing Clarity to the Ambiguous

As CMC's chief financial officer and treasurer, Brian Card's focus is not on production but on revenues, cash flow and other aspects of managing company finances. The human resources department also falls under his general oversight, including implementation of the employee policy manual.

He admitted "doing things right" can be an ambiguous phrase, subject to individual interpretation. "Does it mean to do things correctly, or to have the correct intent? If we're going to do the right thing, we have to follow the policy.

"Doing things right relates to how you allow your employees to act – that's just as important as how you treat the employees. That's why we have codes of conduct, accountability for them to do their jobs properly."

Resolving to do things right may be a commendable intention, Card said, but practicing it must translate into everyday decisions and actions. "There are so many avenues that you have to deal with on a daily basis. Personally, I approach my job with the attitude of being a coworker, rather than a boss. I want to help in facilitating our people to be successful in their jobs, getting their work done properly.

"The company's commitment to doing things right, on a day-to-day basis, must come from the personal qualities of the individuals themselves. We can throw a code of conduct out there and follow it to the letter, but if you don't act according to the spirit of the code, you're still not successful.

"Roy (my father) is not big on telling you how to do things right – he's big on walking the walk. You can observe his actions and he sets the example, showing you what's right and wrong. Talk is cheap. Living life on a day-to-day basis in a consistent manner is the challenge."

Thinking about policies drafted at CMC, Card stressed that application of those guidelines is critical. "The written policy is just a framework. The difficult part involves judgment and reason in how to properly apply the policy.

"Dealing with personnel issues, I try to include compassion and understanding in the process. You can strive to forgive and forget, but you can't ignore or fail to address continual problems when they arise."

Centering on Relationships

Like Brian Card, Mitzi Young sees doing things right primarily in terms of interpersonal relationships. A veteran of more than 25 years at CMC, she works directly with Lewie Card as his administrative assistant and often finds herself lending a listening ear to staff concerns.

"It means a great deal to me that CMC is so family-oriented, that they recognize there are things outside of here that need to be taken care of. Lewie and Charlie both have such an open door policy, if there is something not right, something that I don't feel comfortable with, we can discuss it.

"A lot of times I'm a sounding board for other people. I can tell Lewie and Charlie what I have heard or seen, and they will handle it in such a way that there is a good outcome for everyone concerned. CMC has been built on a Christian-based philosophy that has a lot to do with the open, family-oriented hearts they have," Young said.

"I think our customers greatly appreciate the ethics and values that we have. They appreciate the fact that they are dealing with honest people. We encourage our customers to read our mission and quality statements, so they understand the high standards to which we hold ourselves accountable.

This shows the new CMC Frame Processing Center, which
was a result of its emphasis on Lean Manufacturing.

"And our Values Statement says that if someone believes we are not doing things right, not living up to our values, we want them to let us know. Sometimes our employees will do that directly; other times they may come to me or somebody else that they regard as a sounding board.

"When that happens, the response is not defensive. We strive to make accountability something that applies to everyone, from the top down to the floor supervisor. The tone is set at the top level of management, and that attitude filters throughout the organization."

Plant manager Allen Neely concurred with these appraisals of CMC's people-centered approach to business. "Lewie and Charlie are truly compassionate toward the workers and the workers' families. Not that Lewis and Roy were not, but now it's more openly demonstrated.

"We have been involved in investing in employees in a variety of ways, consulting with them about their finances, about buying houses, assisting them through legal problems, illnesses and other personal crises. The goal of this company is to do everything we can for our employees, and that has been deeply instilled in me from Charlie and Lewie.

"An old school attitude in manufacturing was, 'If you don't like it here, just hit the clock,' but today we realize that if we have gotten to that point, then we all have failed."

Lewie elaborated, noting that if CMC is to attract the most capable people it must seek to retain them in every way possible. "Having top-notch employees has been a big part of our success. Compensation is important, but the way we treat our employees – the way we value them – is more important than anything.

"It goes back to our value statements. Charlie and I admit that we fail many times, but we keep trying. We want to be known as the company that people enjoy working for – to achieve this involves compensation, benefits, the overall working environment, the whole package."

View from the Next Generation

Zach Monroe became a full-time employee at CMC in 2007. Previously he had worked at the company two summers while attending Samford University, earning a degree in business management earlier in 2007.

When initially interviewed for this book, Monroe was in the midst of a management training program. He had worked in the stock room, assembly department, machine shop, engineering and CMC's Creative Tufting Center, had participated in a machine sew-off in a customer's factory, and was starting to get experience in sales. The goal was for him to receive broad experience in all areas of the company. He has since moved into a full-time role in CMC's sales and marketing department.

Doing things right, he observed, "is all wrapped up in the idea that CMC holds the company and its people to a higher standard than the world's standard. I have been around the company all my life, and have known about my dad's philosophy of business. They don't just talk about it. They hold themselves to that standard.

"When you hold yourself and your business to godly standards, it filters down into everything you do – that becomes the basis for why you do things right."

He gave one recent example of the company's commitment to produce top-quality machines down to the smallest detail.

"In the assembly department, I saw a man working to remove two small flaws in a machine guard. We want to fix them so that when the customer receives them, there will not be even the smallest flaws. In every respect, we want them to know they are receiving the high value for which they are paying.

"Doing things right really is a matter of holding to that commitment on a daily basis. The management throughout our plant believes in that philosophy and seeks to carry it out."

Being a member of the family, Monroe noted, does not exempt him from that commitment. If anything, it presents an even greater obligation to uphold the corporate values. "As a trainee, learning and living up to the company values is part of it. It's also the way I was raised – Dad always made sure that we were humble and took nothing for granted.

"I remember hearing him say that if it even *seems* like you are making the wrong decision, don't do it. And Lewie shares that belief. For them, it's not just a matter of whether something is right or wrong, but even whether it could be perceived as wrong. If so, you don't do it. In that way even small business decisions boil down to doing things right."

Newest Addition to the CMC Family

Early in 2010, less than a year out of Auburn University with a finance degree, Lucas Monroe became the newest family member to join the CMC team.

Initially he accepted a position with a financial institution in Chattanooga, but when an offer came to become part of the family business, it was too good to ignore.

"The retail side of banking was not the direction I wanted to go, and then the opportunity to come to CMC came – working with Brian (Card), in the finance side of the business, as well as with my dad and brother, Zach. I couldn't have imagined a better opportunity.

"Growing up I often thought about working in the family business. I always envisioned myself ending up here eventually."

He worked at CMC in the Engineering Department the summer of 2005, prior to college, but his initial exposure to the business came from observing how his father conducted himself, both at home and in a working context.

"I know how my dad is, how he handles things at home – and knew that was how he handled business matters as well.

"It was also clear he knew he wasn't running the business by himself, that he was doing it with God's Word in mind, applying its principles on a consistent, everyday basis. That's what has made this company successful."

CMC has earned a major share of the market for carpet tufting machines, and during his brief time with the company, the company's youngest Monroe has learned why.

"Talking with customers, they have emphasized to me we offer an outstanding, quality product. For instance, a man who has been in our sample department for a couple of years worked with machines from every manufacturer as an operator at Shaw Industries. Not the kind of person to say so just because he works here, he's been adamant that our machines run faster and we make the best-quality machines anywhere."

Monroe fully appreciates the legacy established by the family patriarchs. "Lewis and Papa (Roy Card) passed these values on to Dad and my uncle, Lewie, and they in turn have just continued the family legacy. Our hope is that whoever comes in next, the transition will go smoothly.

"Both Zach and I are still relatively new to the company, but have already talked about wanting it to stay as it is now – even better, if possible. We understand the responsibility of ownership, that God wants us to care greatly about how things will be run in the future."

Seeing Quality as Cause, Not Effect

Even with the tangible benefits of achieving high quality, both in terms of productivity and profitability, from inception this commitment for CMC has been as much philosophical as it has been pragmatic.

"Having a basic desire to achieve Total Quality is part of it, but intention alone can only take you so far," Joyner stated. "Quality is more than just having what you do come out right. It is also about doing the right things, in the right way, at the right cost and in the most productive manner throughout an organization.

"Management must understand that quality has more to do with the way an organization is run than with the condition of their product or service. One is a cause, the other an effect."

Quality products and service – coupled with doing things right – represent lofty business goals that reach beyond the motivations of both short-term and long-term profitability. But for Charlie Monroe and Lewie Card, their motivation transcends even admirable marketplace virtues and values.

To understand the underlying basis for their wholehearted dedication to doing what is right – in serving their employees, customers and the industry – necessitates consideration of a very different perspective on this continuing saga of how tufting machines developed into what they are today.

Tufting-U technical training for CMC customers blends classroom training with hands-on application.

13 BUILDING A STRONG FRAMEWORK

It is not unusual for a company to assign a high priority to quality and excellence, resolving to provide its customers with the best possible products and services. After all, it makes excellent business sense to cultivate satisfied customers that will return to do business again and again. But at CMC, the commitment to the highest achievable standards is rooted even more deeply than that.

Spanning nearly eight decades, the pilgrimage from Cobble Brothers Machine Company to CMC has been both arduous and exhilarating. Since being founded in 1981, CMC has undertaken a unique journey of its own, one that has included a spiritual dimension.

Emblematic of that is a small, unpretentious sign etched into the glass at the main entrance of CMC's facility in Hixson, Tennessee. Just to the lower right of a door leading into the lobby, a simple inscription greets visitors: "Proverbs 3:5, 6." This sign does not beg for attention, but makes a subtle, yet clear statement. It represents a verse from the Bible that reads, *"Trust in the Lord with all your heart; do not depend on your own understanding. Seek His will in all you do, and He will direct your paths."*

Commenting on the sign, Charlie said, "We wanted it there as a reminder every time Lewie and I walk in the door, understanding that it is God's wisdom that we need every day to run the business. If others are encouraged by it in their own walk with God, that's good, too."

He explained how the reference came about: "Lewie and I were general partners together in some real estate. That Scripture reference had been displayed on one of the buildings we had on East Brainerd Road, and we decided to adopt it for our plant when we built the new addition in 2001, following our record sales year in 2000. We were just getting moved in when the events of 9-11 came about. So we think it's fitting that the sign

190

representing words of assurance has been there since then, permanently etched into the glass."

This view outside the lobby at the Card-Monroe Corp. offices in Chattanooga, Tennessee welcomes arriving customers and visitors.

Questions about the biblical reference are "not frequent, but not infrequent," according to Charlie. "Some people comment that they noticed the sign, and those who are familiar with what it signifies tell us they are encouraged by it. Occasionally someone will see it and ask, 'What does that mean?', so we had some business-sized cards printed up, quoting the verse and stating that it comes from the New Living Translation of the Bible."

The intersection of faith, values and business practice also occurs in communications from CMC's leadership to their staff. For instance, one e-mail from Charlie included the following brief prayer, which he said was prompted by reading and meditating on Matthew 7:12: "(God) may we be an organization with important values. May our mission conform to Your will, our work to the Golden Rule, our communications to truth, our treatment of each other to the law of love, our products to integrity, and our accountability to fairness."

Matthew 7:12 is one of the biblical passages that presents the so-called "Golden Rule" – *"So in everything, do to others what you would have them do to you."*

At special observances, such as the company's annual Thanksgiving luncheon, Charlie and Lewie also have acknowledged their belief that God has been very much involved in the day-to-day operations of CMC and express their gratitude through prayer.

Different Paths, Same Destination

Neither Charlie nor Lewie would be described as a "religion on his sleeve" type of person, yet anyone who knows them understands that faith is integral to their lives as businessmen, husbands, fathers and civic leaders. Curiously, the two men embarked on their spiritual journeys from two very different paths.

"My father was the son of a minister, and as a family, it seemed like we were always in church," Charlie said. "My parents, especially my mother – a stay-at-home mom – were godly people. In fact, she was the most Christ-like person that I have known.

"As a child and adolescent, I knew right from wrong and grew up with a certain discipline, although it was primarily self-regulating. It was not until between my junior and senior years of high school that I came to a real understanding about God. That summer, in 1967, I had gone to a week-long Fellowship of Christian Athletes camp in the Black Mountains of North Carolina. It was then that things of God for me moved from my trying to do what was right to experiencing a personal relationship with Jesus Christ.

"One particular night, Bobby Richardson – the former all-star second baseman for the New York Yankees – talked to our group about what it means to really believe in God and follow Him. After he spoke, I think everybody else went to the lodge for ice cream, but I returned to my cabin, climbed onto my bed and after pondering what Mr. Richardson had said, surrendered my life, my efforts – and even my anger – to God, asking Him to take over.

"My anger was the one thing that had always seemed to be getting in the way. It's hard to explain why, but my anger had always been very quick, very much on the surface. Almost immediately it was no longer present, probably before I realized it – but my sister certainly did, since she was sometimes a target of my angry behavior. Some of our neighbors also started wondering, 'What happened to you?'

"But there was another side to my new faith," he recalled. "I was so turned on to the things about God, had become such a different, changed person, that it frightened my father – even though he was a strong believer himself. He warned me, 'You don't want to be fanatical about this.' He was very concerned about my going overboard.

"Overall, it was a really good thing. I found new love for my family, my siblings. God had removed my anger and poured love in its place.

"I remember mowing the lawn and preaching to myself as I was pushing the lawn mower. I was kind of shy and really didn't think I was going to become a preacher, but wanted to be ready if asked. And I did have opportunities to speak at a couple of chapel programs at Tyner High School during my senior year.

"Even so, for a long time I could have been described as what the Bible calls having a zeal for a God, without knowledge. In high school and my years of college, I really did not have much opportunity to grow, to deepen in my faith. My greatest growth as a follower of Christ did not come until after Renee and I had married."

Growing at Card & Co.

Interestingly, the impetus for Charlie's spiritual progress came as much through the workplace as it did from his involvement in a local church.

"When I was doing purchasing for Card & Co., a subsidiary of Tuftco – I started there in 1972 – Fred Holland, a guy representing Siskin Steel, was calling on me regularly. He was being discipled – encouraged in his own spiritual growth – by Ted DeMoss, a very prominent Christian business leader in Chattanooga. Fred and I had many good conversations about spirituality. In addition, Renee and I were attending a church where we received very good teaching about the Bible.

"Then Scott Horn moved into town in 1980 so that one of his children could receive medical care from a local physician. Their friendship with the doctor prompted the Horns to move to Chattanooga. Scott, a former staff member with the Navigators, an international mission organization, established a personal ministry here with his wife, Dolores. I was meeting with him one-on-one; Renee and I were meeting with them and some other couples in their home for Bible study; and the guys (with Scott) and wives (with Dolores) were meeting separately once a week in small groups.

"Scott and his family lived in Chattanooga until 1982, which overlapped the start of CMC, and then they returned to Louisiana, where they were from originally. So for about two years, the Horns had a great impact on Renee and me.

"During that the same time, several men involved with Christian Business Men's Committee (CBMC) came alongside me – Bill Spencer, Ken Johnson and others. So this turned out to be a period of very intense, rapid spiritual growth for me."

Adrift Without a Spiritual Anchor

Unlike Charlie, Lewie had not grown up in a tradition of regularly attending church. So the rules and guidelines for life that he was following at the time he and Charlie decided to become partners in CMC were largely self-made, self-taught.

"In 1981, when CMC started, I was in the middle of a troublesome downhill slide," Lewie recalled. "I was living my life in a way that I thought was right – drinking, drugs, and all the activities that went along with that.

"Being around Charlie, seeing somebody that didn't do all of those things and observing the peace he had that I did not experience in my own life, made an impression."

Lewie described his life during those years as, "having everything (materially) that I could ever imagine – but being more miserable than I could ever imagine as well.

"Then in 1987 I was introduced to Roger Erickson by a friend. He was an executive with CBMC at the time and eventually became a very strong spiritual influence in my life.

"I remember when I met him, one of the first questions he asked me was, 'If you were to die today, where would you go?' I had the textbook answers – I said that I felt I was a good guy, gave some of my money away to good causes, and (sometimes) went to church. Roger explained that although those were good things, they were not what will get you into heaven.

"I spent two years meeting with him in what he called a discipleship relationship – and all the while I was still maintaining a crazy lifestyle. I was bent on trying to do it my way. Through reading the Bible and learning what it says about God and Jesus, I had accepted Christ in my head, but had been unwilling to let go of my lifestyle. I just kept doing things the same old way I had been accustomed to – the wrong way.

Reaching a Turning Point

"My turning point came in September of 1989, when I ended up in a rehab center in Smyrna, Georgia for drugs and alcohol. I remember getting on my knees there and finally surrendering my life to Christ. Things in my life had felt like they were starting to implode, and I finally was willing to admit that everything was out of control. I called out to God, and from the very moment I said that prayer, asking Jesus to come into my life, I knew He had accepted me. I felt my burdens – what I now refer to as the trials and tribulations of my life – lifted off my shoulders at that time.

"The reality was that the world had not changed. My circumstances were the same. But God had begun the process of changing me from the inside. Since that date in September, nearly 20 years ago, I have no longer had any problems with alcohol, drugs, or chasing after women. And that's just the beginning – life has gotten better and better ever since."

Charlie said there were obvious changes in Lewie's life after he turned control of it over to God, but not necessarily as dramatic as one might think.

"Lewie was always very receptive to everything I said or even suggested, when it came down to a business decision of great importance. So I never felt, even before he became a follower of Christ, that there were any principles or decisions that we needed to compromise."

As Lewie has termed it, "I really was a good guy, but I was going to hell."

"One thing I have always been impressed with about Lewie, and his whole family," Charlie commented, "is there is no jealousy, no sense of competition – even between Lewis and Roy. So Lewie and I may not always agree or see eye to eye, but there has never been any jealousy. What a freeing thing that is – it's a huge part of what has made our relationship work."

A Changed Outlook

Lewie explained that at work, "I always had an eye on the end result. That's not to say I did not have any inner conflict, but my growth in the Lord – my faith and trust in Him – made a tremendous difference.

"One thing that helped me a lot, whenever I did encounter internal battles, was that Christ had set in my mind the importance of submitting to authority. Especially to His authority. I did not have to compete for that responsibility.

"Someone once used the analogy of a corral to explain this. When you are confined in a corral, you know how far you can go. When you're not a believer, you don't have that corral and you can go too far."

After Lewie turned his life over to Christ, several positive developments resulted, according to Charlie. "Then we truly became equally yoked as business partners. It showed in the unity that we had. We started pulling together, not only professionally but also spiritually. We started to attend a men's Bible study group together that was led by Bill Spencer and Dave Nabors. Bill helped us to learn what it meant to abide in God's peace, and Dave helped us to realize what it meant to truly love your wife in a sacrificial way. We were able to observe genuine, godly men as they served as examples of the Christian life by the way they felt and acted, and these men also provided accountability for us."

The huge Promise Keepers gatherings in the early 1990s provided additional stimulus for spiritual growth. Charlie and his good friend, Bill Riley, took Lewie to the second such event ever held, in Boulder, Colorado, and that served as a time of spiritual awakening for them. "Being there with thousands of other men who loved Jesus was a big step in our journey of faith, seeing that you could be a man and love God, too – that it was okay to be open about what you believed."

They later attended another mammoth Promise Keepers gathering in Atlanta, where more than 40,000 attendees were challenged to "find a man of color in your life and reconcile with him," Lewie recalled. "The passion that I have today for people in the inner city has come from that."

Signs of a Transformed Life

Today, along with CMC's Mission Statement, Values Statement and Quality Policy, three other documents are displayed prominently in Lewie's office. Each reflects his new outlook on life, one that has become centered on God's purposes rather than his own.

One statement posted on the wall is what he refers to as his "life verse," Matthew 25:40, which states, *"The King will reply, 'I tell you the truth, whatever you did for one of the least of these brothers of mine, you did for me."*

"This has become my passion," Lewie said. "God has led me to help others, and it describes how I want to spend my life."

The second statement is another mission statement, except this one he has adopted personally:

"To let God control every part of my life from the time I wake and through eternity. With Christ in me to help others to know the peace found in putting Jesus Christ first in their lives, no matter where I am or what I'm doing."

Lewie explained that in a book he had read, the author suggested it would be useful to develop a personal mission statement, "so I did it to keep my compass in the right direction – as a reminder."

The third sign posted in his office was taken from the Jan. 9 reading in a daily devotional book, *Daily With the King,* by W. Glyn Evans, that he found especially meaningful several years ago. It describes what Lewie now regards as the definition of genuine success:

"True obedience is always successful. It always accomplishes God's will. It is immaterial to me whether my cancer is cured, my promotion achieved, or my loved one saved. If I have obeyed, that is enough. I am successful. Beyond that, it is God's business. If I insist on a happy, selfish outcome, then I am meddling. The instant I obey, success is automatic – God's success, not mine."

Applying Principles to Business

Attending a Business By the Book seminar, sponsored in Chattanooga by CBMC in 1989, was another spiritual milestone for both Charlie and Lewie. "It was a huge step in helping Lewie and me get on the same page in our desire to run our company according to sound, practical biblical principles," Charlie noted. "It also gave us a basis for establishing criteria as we considered other people that wanted to go into business with us. We determined that everyone we partnered with would be required to listen to the Business By the Book tapes and agree to the principles they teach."

In the mid-1980s, Charlie became part of a long-distance accountability group with several accomplished business leaders, including Ken Johnson of Chattanooga, Ron Huber of Phoenix, Arizona, Don Mitchell of Detroit, Michigan, Albert Diepeveen of Kankakee, Illinois, and Dick Nelson of Minneapolis, Minnesota.

"This was really good for me," Charlie said. "I felt a little bit out of place with those guys, since they were so much more experienced and CMC was just a little growing company at the time, but they embraced me, welcoming me into the group.

"I went to Phoenix to attend Ron's Time:Systems time management seminar and found it had direct application for our business. In fact, our biggest customer, Shaw Industries, had been using the program extensively through their organization as well.

"It was through Don Mitchell that I was introduced to the Total Quality Management philosophy, long before we met Jim Joyner. Don was heading up Total Quality Management for General Motors at the time and knew W. Edwards Deming, who had pioneered this approach.

"Albert Diepeveen had a huge heart for giving money to charitable causes and being generous, and I learned a lot from him. Ken Johnson was a tremendous networker among business people and a great encourager. Even today, I think there are 500 guys out there that would say that Ken is their best friend.

"He started the Christian Network Teams ministry, enabling business owners to come together on a monthly basis for mutual encouragement, problem-solving, accountability and support in living out their faith through their companies. Lewie and I eventually became a part of CNT locally, in different groups. I already had my accountability group that we continued to maintain primarily by phone, but CNT gave it a more local flavor.

"It's been good to be able to benefit from seeing what leaders in other local companies are doing to live out their faith on a daily basis."

"Through my involvement in CNT, I discovered a couple of businesses led by men I greatly admired in how they did things to treat their employees properly," Lewie noted. "Among these have been Chuck Mills and Keith Eischeid at Steward, Inc., whom I had met through my early involvement with CBMC, and Chuck Zeiser and his sons, John and Bruce, at Southern Champion Tray.

"They had well-established, older companies. I determined that when CMC grew up, we wanted to be just like them. Especially with Chuck Zeiser, it was an opportunity to understand our calling here – as a ministry within the walls of a business – and seeing enduring relationships come about as a result.

"Ken Johnson has always pointed out that men he knew would hurry up and try to get their business done, so they could go out and do ministry. CNT has taught me that in reality, being in business *is* ministry if you approach it in the right way."

Charles Monroe (left) is President of Card-Monroe Corp., and Lewis Card, Jr. is Vice Chairman and Secretary.

Adopting an Eternal Perspective

In working together and learning to integrate sound biblical principles in their business, Charlie and Lewie discovered the value of operating according to a perspective that is not just long-term, but eternal.

"Having an eternal perspective affects your view of ownership," Charlie explained. "Whether you want to admit it or not, God owns everything, even though shareholders may think and act otherwise. When there is a downturn in business, or you're facing decisions that no one wants to make but are so consuming, the reality is that God may be trying to pry our fingers away from holding onto the business. Situations like that remind me to open my hands and remember that ultimately, it's His business. God can give it – and He can take it away. So we need to think seriously about questions like, 'Whose is this?' 'What's it all about?' and 'Why are we doing this?'"

Lewie agreed: "The Bible teaches that we do have a way out, a scapegoat – Jesus died on the cross for us. I have been through downturns of business; a divorce; having my wife, Margaret, struggle with terminal cancer and ultimately be called home to be with her Lord; challenges as a parent – a tremendous amount of extremely difficult situations. But through it all, I know God has a plan. So I don't have to be anxious about it. His plan is much greater than mine. I am here to honor Him and be obedient to Him. When I am able to do that, that's when I am truly successful. And to be honest, I have learned that it doesn't make sense to be otherwise.

"I like to look at it this way: Think back to the biggest problem you had in 1987, 1992, or even 2005. What was it that kept you awake and seemed to be consuming all of your time? Most of us probably can't even remember what those things were. That tells you how important they really were.

"I would rather have God's worst than man's best – I can tell you that for sure."

A special day early in 2010 marked another example for Lewie of how God can bring blessings out of sorrow. A few months after Margaret had passed away, he had begun dating Rebecca "Becky" Lee Hatfield, a well-known dentist in Dunlap, Tennessee. They were married on January 16 at New City Fellowship in Chattanooga.

"In the time it takes to write a book, from inception to publication – in the case of this book, *Tufting Legacies,* several years – we experience many important life changes," Lewie observed. "God's bringing Becky into my life after Margaret's death was one of them, and I'm so thankful for how the Lord continues to bless me even through tragedy."

Shaping Decisions and Relationships

This ultimate long-range view also has strongly influenced how they approach everyday corporate decisions and relationships with customers, suppliers and employees.

"Just as Lewie and I, in our own spiritual walk, are trying to keep an eternal perspective in our personal lives, we also try to keep a long-range perspective in business – for instance, making sure decisions that we make are not knee-jerk reactions, and at the same time keeping our long-term goals in mind," Charlie noted. "The fact that short-term pain is sometimes necessary for long-term gain is OK with us.

"For example, one of the benefits of a closely held private company is that your shareholders have a good understanding of the business. We don't make decisions based on the stock price of today. We're trying to build long-term value for the good of everybody involved in CMC. Unlike what you sometimes see with publicly traded companies, we don't optimize for today and sacrifice the future."

Charles Monroe (standing in back) speaks to CMC personnel at
the company's annual Thanksgiving luncheon in 2007.

"This perspective reinforces my realization that it's not about me, but about everybody else – customers, suppliers, employees," added Lewie. "It's not about me winning and being happy. I may get those results, and they may actually be better, when I think about and put others first. But that is the byproduct, even if it's a very good byproduct, not my primary objective."

"It goes back to how we value each person as an eternal being. In light of eternity, just how important is this issue that's keeping me awake at night?" Charlie explained.

"We both believe that one day we will have to answer to God – give an account – for everything we have done. I want to have a good and pure heart, to have a good answer for my actions," Lewie stated. "Personally I am very competitive and that shows up in a lot of ways. I want to win all the time, but having that attitude can hinder you in doing what's right. Spiritual values serve as a compass to counterbalance potentially destructive attitudes."

Words Carried Out in Actions

Consultant Jack Hatcher commented on how the eternal perspective maintained by Charlie and Lewie manifests itself in the company's daily operations.

"At CMC, Charlie and Lewie and others on their team exercise the long view. That is not at all to imply that they are not profit-conscious. They manage the company well, and thoroughly understand the dynamics of business. They know that more complex machines have greater value than simpler machines, and understand how costing is improved with increased utilization of manufacturing capacity. They recognize where manufacturing efficiencies marry with product mix and manage that equation well.

"So they are not discounting the importance of profit. But in their scheme of things, profit is of secondary priority. That is very atypical, even rare – out of this world, really.

"Most companies are in business primarily to maximize shareholder equity – that is why they exist, and they are programmed to do so annually, if not on a quarterly basis."

The fact that CMC is privately held and debt-free is advantageous to its desire to live out its commitments, according to Hatcher. "Its leadership is both intelligent enough and courageous enough to insist that the business be performed correctly, assuring complete customer satisfaction. They are willing to accept a down year if that is what's necessary to fulfill the commitment it has to its mission, vision and quality standards.

"Rather than seeking to placate shareholders, they opt for satisfying the customer every time. In my 14 years of consulting with them, I have never seen that not to be the case."

'Ethical, Honest Men of Integrity'

Jim Joyner offered similar observations about CMC, drawing from his 40 years in dealing with dozens of companies around the world, both as a consultant and a full-time employee.

"It is so good to work in a company with ethical, honest men of integrity at the helm. Their values permeate the business and shape the direction it takes. We frequently talk about values here. It is common for us to take our values statements off the wall and refer to them, using them as a constant reminder of what the company is about.

"This is as well-run a business as you will ever find. I have been privileged to work for other companies that have all been well-run, and I have sat in the board rooms of hundreds of companies – GM, Sony, Johnson & Johnson, Upjohn, Ingersoll Rand and others – but I have never, ever encountered an organization that is any better than Card-Monroe.

"Charlie and Lewie both have a great sense of the heritage that the company has evolved from, and both are great caretakers of what has made the company good and successful. They have been both very constant in leading and guiding the organization. Their consistency has never been in doubt – and that is a mark of all good leaders."

Articulating Their Faith

Neither Charlie nor Lewie is reluctant to verbalize their values and beliefs that undergird them. However, they recognize the importance of consistently practicing the convictions that they profess.

"I like the quote from Saint Augustine," Lewie said. "He said something to the effect, 'When testifying about your faith, if all else fails, use words.' We need to be consistent in how we live and conduct ourselves, how we interact with others. I hope that when people watch me, they will see a difference, especially if they were to compare that to my old self, or compared to what we see a lot in everyday life.

"It was industrial leader Wayne Alderson who said, 'People don't care what you think; they want to think that you care.' And that is our goal and desire. As far as our customers are concerned, we would hope they see a difference in how the company is run, as well as in the caliber of people who work for us."

A CMC Infinity machine manufacturing carpet tile in a Georgia mill.

Charlie agreed: "I hope our employees feel accepted, regardless of what their views are, their perspectives, where they are coming from – no matter where they are spiritually, or aren't. That's my hope, that they sense acceptance and understand how important they are to us.

"In our company we don't have anything religious – we don't hold church. We simply have a heart for our people, which we believe reflects God's heart for us. That's one of the most difficult things to communicate. Sometimes I feel we aren't doing a very good job of erasing 'we/they' thinking.

"Our desire is to help people going through personal difficulties. We may not always be able to give them the time and attention they deserve at a particular moment. But over the past number of years, we have had someone available to offer answers, to build trust among our employees, to provide the direction and counsel that they need.

"In our employee assistance program we have an individual available, on location, one day a week. He spends part of the day here and part of a day at our Jackson Plant. But he is available 24/7 for employees in crisis. We have had a number of situations within the company, such as a family member attempting suicide, a tragic accident, the death of a child or a spouse, as well as various life issues – emotional concerns, financial struggles. Lewie and I, and our management team, don't always have the time or opportunity, much less the knowledge, to deal with such circumstances. So it is a great help to have someone on hand who knows how to respond when needed.

"But it's more than just dealing with clinical needs. It also involves relationships, building mutual trust with our employees. He walks around the plant, asking how each person is doing, asking about their families, just getting to know them. We want them to know that someone is available to them, ready to help if needed."

Others-First Emphasis

While they hope their employees will appreciate the spiritual foundation that guides their company, Charlie and Lewie affirmed that embracing the same beliefs is not a requirement, or even an expectation, for employment at CMC. Federal employment laws prohibit such stipulations, but the leaders also understand that spiritual convictions are an individual matter that cannot, and should not, be forced on anyone. Instead, they look for people of character who not only have individual skills but also fit CMC's others-first culture.

A CMC staff member at his work station, designing a tufting machine part using advanced 3-D modeling software.

"We try to discern just how relational they are," Charlie asserted. "Is a job candidate someone that is able to lay down personal agendas for the good of the whole? We want skilled and talented people, but for the most part, we want people that are willing to work together as part of a productive team.

"Giftedness, education, abilities and skill sets are important, but if people can't function together well as a team and fit into our culture, they can still be destructive.

"Do you have to be a Christian to work here? No. We enjoy having people come here and embark on their own spiritual journeys, wherever those may lead them. But if they have worked at other places previously, when they come here we would hope that they see the reality of how special the people are that work at CMC and would be excited about joining us."

Stewards of Resources and Responsibilities

Does the faith Charlie and Lewie share make them better executives, better business people than peers who operate according to different spiritual frameworks – or with no spiritual reference point at all? Not necessarily, but it is clear that they perceive themselves not so much as owners of CMC, but rather as stewards, persons entrusted with significant resources and responsibilities for serving those who work for them, the customers who buy from them, and the community in which their company is situated.

In the Bible, Jesus stated, *"From everyone who has been given much, much will be demanded"* (or, *"to whom much is given, much is required"* – Luke 12:48), and at another time observed, *"What good is it for a man to gain the whole world, yet forfeit his soul?"* (Mark 8:36). The apostle Paul, writing to the thriving church in the city of Corinth, stated, *"Now it is required that those who have been given a trust must prove faithful"* (or, *"It is required of stewards that they be found faithful"* – 1 Corinthians 4:2). Lewie and Charlie do not take admonitions such as these lightly, understanding they are accountable for how they conduct themselves and operate their business.

Two other Bible passages describe the unusual spiritual bond these corporate leaders share. One comes from Proverbs 27:17, which states, *"As iron sharpens iron, so one man sharpens another."* The other is from Ecclesiastes 4:9, which affirms, *"Two are better than one, because they have a good return for their work."* In addition to complementing each other personally, Charlie and Lewie have supported one another spiritually, seeking to build and sustain a company committed to excellence, quality and values that ultimately point "all to the glory of God."

These convictions will continue to guide their corporate planning and management philosophies as they look to the future. Their thoughts about what may lie ahead in years to come – for CMC specifically and the industry as a whole – comprise the next and concluding chapter.

Plaques commemorating dozens of tufting machine patents by Lewis and Roy Card, and others at CMC, adorn hallways at the company headquarters.

14 WHAT LIES AHEAD IN THE FUTURE?

On August 2, 1992, President George Bush was touring the facilities at Shaw Industries in Dalton, speaking with executives, employees and civic leaders. "In the history of your industry, you find a parable of American progress," he said. "It starts simply – families selling hand-tufted bedspreads that they made themselves on Highway 41, Peacock Alley, and it continues with these sprawling factories that sprung up after the war, rolling their carpets into homes and offices in every corner of America."

What then-President Bush failed to acknowledge were the largely unsung contributions of people like Joe and Albert Cobble, Lewis and Roy Card, and many others that have been cited throughout this book. Without the imagination and ingenuity of true pioneers like the Cobbles and Cards, the humble tufting experiments of Catherine Evans Whitener might not have progressed far beyond relatively crude bed coverings. Today most of us might still be spending our days at work and at home walking and standing on hardwood, tile or linoleum floors, rather than soft, plush carpet.

The concluding lines of this "parable" remain to be written as companies like CMC continue refining and enhancing the already vast capabilities of tufting machines, and as mills like Shaw, Mohawk, Beaulieu and others strive to offer still more distinctive looks and styles for carpet buyers to consider.

In his book, *Bedspreads to Broadloom,* Thomas M. Deaton noted that although the basic principles and processes of tufting have long been established and remain fundamental to the industry, there has always been room for innovation and imagination. "This is still an industry in which hard work, creativity and inventiveness pay off in dollars… The tufted carpet industry is still the American dream at work."

Indeed, nearly six decades after the first crudely designed tufted carpet rolled off their machines, no one believes that the final words have been pronounced about cut or loop pile, or any combination of the two.

There have been no greater authorities on the fantastic journey that tufting has undertaken than its patriarchs, Lewis and Roy Card. More than anyone else, they have been able to appreciate – and yet marvel at – how far the industry has come.

From One Needle to More Than 2,600

Reflecting back over the nearly 70 years he has been involved in the tufting business, Lewis observed, "We came from one needle to more than 2,600 needles in tufting machines. The developmental work was very exciting – when you finally got it developed. It was a tough business, but also extremely interesting. I and my family have done very well because of it. Those tufting machines still amaze me – the amount of goods they can produce!"

He recalled the efforts to devise a means for controlling the first scroll machine in the early 1950s. "We had a pattern cut into the surface of a drum that would allow metal fingers either to make electrical contact with the metal portion of the drum or to be insulated from the drum and breaking contact, thus controlling the yarn feed on the machine. Many years later, computers do the same thing today, only in a much faster, far more sophisticated way.

"It's amazing what the computer has done (for tufting machines) – but that's probably true for a lot of machinery. At our Jackson Plant in Ringgold, Georgia, they are now making parts in eight hours that took us three weeks to do in the old days. And with the help of computerized machines, we're doing things that we never would have thought of before."

An electrical technician at CMC works in the Electrical Production Center for Tomorrow today (EPCOTt) assembling a machine controller.

Roy added his affirmation to the wonder of how far the industry had progressed during their lifetimes. "Today when I go to the plant and see what they're doing in the sample and development department, it just amazes me. Some of the designs you are able to do today, the ways you are able to plant yarn in the carpet, seem incredible to me, even though we came up through the earliest days of the industry and spent so much time trying to figure out all of that stuff.

"There is very little woven carpet in the United States anymore, and it has almost gotten to the point where you can't look at carpet and tell whether it is woven or tufted, because of the advances in tufting quality, design and styling."

Curiosity always was a motivating factor for Roy in his pursuit of innovations, and he said that has not changed in his latter years.

"When I go somewhere, I still like to try and find a corner so I can see whether the carpet there is tufted or woven – you can't always tell. But I find myself paying attention to the carpet wherever I go."

Seeing the growth of the carpet industry into a multi-billion dollar, global enterprise, Roy admitted that was not a development that he and Lewis ever envisioned. "During the early days, most people saw how little carpet was being sold when it was being woven, so they were not too enthused about it. Neither Lewis nor I sat down and thought that one day 50 million square yards of carpet would be produced. But every year the amount of carpet tufted went up – it became better looking, better quality, made faster and cheaper."

Inventor of the Year

Over the years the Cards have received occasional recognition to their copious contributions to the industry, although sometimes even that came with conditions attached.

For example, in February 1986 Roy was honored as the Chattanooga Inventor of the Year by the Chattanooga Engineers Club – the first year the award was given. He was selected for his invention of the cut-loop process and apparatus for carpet tufting machines. Effectively, this device enabled machines to make a high cut-low loop pile for a sculptured pattern effect on carpet.

The notification letter read, "Criteria established for the recipient of this award is that of an inventive contribution which impacted substantially upon the general economic development of the Chattanooga area. Your invention clearly did that."

Apparently, however, the impact on the "general economic development" of the region could only extend so far. The $16 per person dinner cost even applied to the award-winning Card and his wife, who paid for their own meals!

Hall of Fame Inductees

In recognition of their contributions to "Chattanooga's rich entrepreneurial spirit," both Lewis and Roy Card were honored by the University of Tennessee-Chattanooga by being inducted in April 2010 into its Entrepreneurship Hall of Fame.

A press release for the occasion, from the university's College of Business, described the brothers as "true pioneers, blazing the trail for today's multibillion dollar, global carpet industry.

"Over the years these men started and acquired a succession of tufting machine companies, including Super Tufter, Card & Co. Inc., Southern Machine Co., Tuftco Corp., all in the Chattanooga area, along with Gowin-Card in Dalton, Cobble Brothers, Ltd. (England) and British Tufting Machine (England), before joining their son and nephew, Lewis Card, Jr., and son-in-law, Charles Monroe, at Card-Monroe Corp. Together they hold more than 100 U.S. and international patents for tufting technology."

In addition to celebrating the area's rich spirit of entrepreneurism, UTC's Entrepreneurship Hall of Fame was created to honor entrepreneurs and present their success stories as a role model for business students at the university.

Impact on a Region, a Nation, and the World

Unlike his brother, Lewie, Lamar Card chose not to get involved in the tufting industry. However, he has long remained a proud observer of his family's achievements with tufting machines. He commented on the impact the Card brothers' tufting developments had, not only on the carpet industry specifically, but also on regional, national and international economies – "how the availability of solutions for flooring ultimately affected the lives of so many people on a certain level," whether in the Chattanooga and northwestern Georgia area, across the United States, or eventually, overseas.

"About a year ago, I was in rural Moscow, Russia. The economy there was exactly where the United States was in the early 1960s. If you were to drive through Moscow today, you would observe a sea of what we typically call condominium complexes. Part of this phenomenon is a shift from wooden, Persian rug-covered or cement floors to tufted carpet. They were constructing apartment buildings, planning to put floor coverings into the residential units.

Suddenly there was a demand for tufting, and companies were being started up to meet that need.

"My dad and his brother influenced generations in this area, the nation and the world. It was during the post-industrial period that their inventions started taking hold on a national and international scale. This work spawned entirely different industries, most notably the tufted carpet-producing business.

"My dad's work was highly specific, but it had a much broader cultural impact. Nationwide, beginning in the 1950s, an entire society switched from one type of stuff below their feet to something very different. Culturally, I would compare that impact to changes in the types of hats people wore, or the kinds of cars that they drove."

Roy's son Brad agreed, observing that despite their accomplishments, the brothers managed to remain remarkably low-key and self-effacing. "Their influence in the tufting industry is undeniable, whether you're in Belgium, South Africa, Russia, or wherever. The technology and machinery for making carpet has resulted from their endeavors. Lewis and my father, and CMC have had a very real impact on the world – that's something we can be proud of. The amazing thing is that it's all been done in a quiet, humble manner. My dad is a very humble man, never one to brag on himself."

Added Lamar, "Dad was at the heart of all this from the beginning. He showed Joe Cobble how this could happen, and shared his vision with others."

Genealogy has been a favorite hobby for Lamar, and he said the Cards' ethnic heritage might have played a contributing role in the brothers' achievements. "I don't think personality traits are genetic, but I think there definitely are some common personality characteristics among the Scots-Irish, traits that my father and Roy Card exhibited."

Some of those, he said, include being extremely frugal, stoic and tenacious. Since Lewis and Roy are from both Scots-Irish and German (Cobble) stock, these traits could have been derived from both sides of the family. "Dad is perhaps the most successful entrepreneur of the 19 Card generations that we have traced."

The Cards, who meet annually for a family reunion in Soddy-Daisy, Tennessee (about 200 people attended the event in 2007), have traced their lineage back to Somerset, England and the county of Dorchestershire in 1521. Direct descendants of the Cards, the family historians have learned, go back to Southern England and the London area.

Lamar Card's observations find support in *Born Fighting: How the Scots-Irish Shaped America,* by James Webb. In his book, Webb describes the Scots-Irish as a resolute people of "insistent independence, willingness to fight, and refusal to submit" that influenced much of what is now known as Great Britain, as well the United States when settlers immigrated to the "New World."

"The measure of a man was not how much money he made or how much land he held, but whether he was bold – often to the point of recklessness – whether he would fight, and whether he could lead…," Webb writes. "Physical courage fueled this culture, and an adamant independence marked its daily life. Success itself was usually defined in personal reputation rather than worldly goods. …

"The famed Scottish talent for inventiveness and adaptability showed itself on the frontier regions again and again…the courage and innovative talents of its soldiers and pioneers as well as the evolution of an adamantly independent style of democracy. …

Webb adds, "Our fathers were not scientists … they were doers, fixers, mechanical geniuses, risk-takers … ."

While neither Lewis nor Roy, unlike their ancestors, became known for any military prowess, they still exhibited a fighting spirit in confronting difficult challenges and adversity, as well as the day-to-day battles of keeping a business running. And their inventiveness certainly exhibited a high degree of mechanical savvy.

The CMC Board of Directors includes (seated, from left) Lewis Card, Sr., Janice Card Henderson, Roy T. Card, and standing, Lewie Card, Brian Card and Charles Monroe in this 2006 portrait.

Responding to some who have conjectured that Lewis's and Roy's lack of formal engineering training may have kept their thought processes natural and untainted, Lamar said he believed that even with formal training, Lewis would have retained his intense, highly focused personality and perhaps would have been even more successful – in tufting or any other field he had gone into. "Dad, with training, even could have become an excellent experimental psychologist," said Lamar, who has studied psychology himself.

"They never formalized, as a lot of companies do, the structure of R&D (research and development) in the companies Dad led or started. In a sense, he and Roy *were* the R&D department."

The Benefit of Experience

Today CMC stands as a direct beneficiary of the expertise that Lewis and Roy acquired through those years of diligence, fueled by their never-quit spirit. CMC built upon the rich heritage of their innovations and, particularly early on, was able to capitalize on their wisdom and insights firsthand.

When Lewis and Roy became a part of CMC in 1983, this provided an opportunity for Lewie and Charlie to concentrate on the everyday operations of the business, while the brothers resumed their focus on development and innovation.

Even though Charlie and Lewie, along with Brian and Brad Card, have many years remaining in their careers, CMC already is seeing the arrival of another family generation in the business.

"With Keith Askew, Ryan Berube, Zach Monroe and Lucas Monroe, the fourth generation is just starting to come in," Charlie noted, "and we hope they will learn the business and be able to continue building on the foundation that was laid by Joe Cobble, Lewis and Roy.

"Of course, CMC has always been more than a family enterprise. Today we have many talented people here, far too many to name, with no connection with the Card family that play extremely significant roles in what we do. But it is also exciting to see that some of the second generation still has remained involved – Roy at 80 and Lewis at 89."

What's Ahead for the Future?

For those of us who have grown accustomed to carpet being a standard feature of most domestic and commercial settings, it's hard to comprehend how far the craft of making carpet has come since the early 1900s. It has changed from a household luxury that was the exclusive domain of privileged society to a staple of virtually every home in the United States – and increasingly, in many households throughout the developed world. Carpet also commands a prominent place in workplaces, educational institutions, care centers, shopping malls, restaurants, hotels and motels.

The Carpet and Rug Institute likes to emphasize, "carpet creates a better environment…it provides an important measure of safety and comfort in rehabilitation facilities; creates a calmer, quieter atmosphere for young children to learn; makes homes warmer and more attractive, offices more productive, and retail settings more inviting," and today, tufted synthetic sports fields commonly enhance the competitive experience.

But what is the future of carpet – and the industry as a whole?

"Even though the industry has advanced incredibly since the 1950s, we still are coming up with new ideas," Charlie said. "In fact, we have adopted a new corporate brand identity statement that we believe fits us very well – 'Creating Possibilities.'

"Recently CMC's Wilton Hall, along with Bill Christman, developed a revolutionary new technology, finally achieving something that has only been possible through the weaving process. We have named it ColorPoint and think it will eventually revolutionize the industry."

ColorPoint enables carpet manufacturers to create Axminster woven looks in tufted carpets, offering unprecedented styling flexibility, along with cost-savings for the consumer.

"Our new ColorPoint technology makes it possible to place up to six colors or more anywhere in a carpet pattern with unmatched pinpoint accuracy, with no buried ends, to provide full-gauge coverage," Charlie stated. "The new machine utilizes the Infinity Pattern Attachment that helps create carpet with multiple pile heights and a variety of textures and cut-loop effects, and there is no limit to the size and scale of patterns that can be created in the design."

There are three models of this innovation – ColorPoint LP, an all-loop pile machine that can produce Axminster-type multicolor patterns; ColorPoint CP, an all cut-pile multicolor pattern machine that can place up to four colors anywhere in the carpet design; and ColorPoint C&L, which can produce cut-pile multicolor patterns or patterns with varying combinations of cut and loop.

"This is groundbreaking technology, a huge development," he said, allowing carpet manufacturers to use a wide range of colors in a carpet design without having to bury or hide shorter loops of other colors. "Interior designers surveyed have said they wanted a new array of patterns, without the linear appearance that has been a limitation of older, existing technology.

These carpet samples are just a few examples of the capabilities of
CMC's revolutionary new ColorPoint tufting technology.

"These machines can produce gracious, beautiful non-directional carpet
in very subtle colors and patterns for residential purposes, as well as the large,
bold designs you often see in hospitality installations. Today, patterned carpet
is what designers are looking for and it's the best selling, so we are especially
excited to provide our customers innovative, value-added capabilities to
stimulate market interest."

Charlie stated that although the recent recession initially slowed the
introduction of the ColorPoint technology into the industry, interest has
grown along with the resurgent economy. "We've had a very strong, positive
response to ColorPoint. Customers will have at least eight of the new machines
in operation before the end of the year, and they're very excited about the
many benefits this cutting-edge technology has to offer."

CMC's new ColorPoint technology has opened up carpet tufting possibilities more than ever before, reproducing craftsmanship previously possible only by weaving.

Always Looking for Improvement

While ColorPoint is perhaps CMC's most dramatic development, Charlie knows it will not be the last. "We are always seeking to improve our technologies, and to learn how we can utilize what we already know in better, different ways. Every CMC machine is designed to capture the manufacturer's artistic vision and transform the genius of a moment into a dependable tool for ongoing success. Our new ColorPoint series of machines is just the latest in our long line of innovative solutions for the carpet industry.

"Essentially, carpet is a fashion industry," Charlie pointed out. "We are giving the mills equipment that enables them to supply their customers with styles that are fresh and new, that they have not seen before; to continue to be low-cost producers through greater speed and efficiency, and to make products of even better quality."

"Our corporate mission statement calls for us to be the tufting equipment supplier of choice throughout the world," Lewie said. "This is helping us to achieve that goal. We are constantly learning and always seeking to discover what is new out there that we can apply to our machines, as well as attending trade shows just to keep current with advances being made."

As this book was being written, CMC was making its largest-ever capital expenditure, investing in new equipment for precision main-frame machining for its tufters and erecting a new building to house the equipment. This machinery will be used to machine all of the main frames, from the 42-inch sample machines to the 204-inch carpet machines that produce five-meter or 17-foot wide carpet.

CMC's Infinity attachment features servomotors that control the flow of individual yarns to offer an ever-expanding array of carpet designs and patterns.

"We are always updating our manufacturing equipment. We're never satisfied with status quo. We try to take the long view, rather than focusing on the immediate return," Lewie said. "In terms of equipment, we buy the best equipment to utilize in our own machine shop for our machines, just as we hope our customers will buy the best equipment (ours) for many years to come.

"When it comes to dealing with customers, we try to look at the relationship 20-30 years down the road, not just to making money today. In dealing with people, you have to look long range if you have their best interests at heart. Improvements are not cheap, but we are committed to reinvesting in the business for the long haul."

Tufting for Athletic Facilities

In recent years, one of the new directions for tufting has been the manufacture of artificial turf, using a process basically the same as tufting residential or commercial carpet. The idea was initially conceived in the 1960s, when domed sports stadiums were first being built. The problem with Astroturf and similar products of that time was that although they looked like grass from a distance, they were not nearly as resilient or kind to the human body on impact. Over time, because its concrete-like hardness and sandpaper-like abrasiveness caused injuries both major and minor, artificial grass suffered a falling from grace.

An aerial view of Chattanooga's Finley Stadium, with
artificial turf that was tufted on a CMC machine.

By being able to tuft higher pile artificial turf, the surface has a more natural, grass-like feel and effect. As a result, artificial turf is enjoying a phenomenal resurgence in popularity. There are more than 700 turf fields in the United States this year, and they are being installed by the thousands in Europe, China and other parts of the world. In their hometown of Chattanooga, fields made with CMC's turf tufting machine have been installed at Finley Stadium, home for the University of Tennessee at Chattanooga football team, and stadiums at The McCallie School and Baylor School.

Russ Huesman, head football coach at the University of Tennessee-Chattanooga, stands in front of the tufted grass field at Finley Stadium, the home of the Mocs.

Artificial turf has special appeal, as well, for golf courses with sandy soil, even being used for greens and fairways in some cases.

This, too, is a development that took Lewis Card by surprise. "I never did think of that as an application of the tufting process, but there really is no problem in tufting artificial turf when you have the demand for it. Theoretically speaking, you can tuft any yarn or material that can pass through the eye of a needle.

"When Astroturf was first developed with synthetic yarns for Houston's Astrodome because they could not get natural grass to grow indoors, it was very low and thin. Due to the fact it was placed over a hard asphalt base, player injuries were common. In response, the artificial turf industry over time reinvented itself. Instead of producing a grass pile of a half-inch or less, it's now tufted with three-inch pile that is in-filled with sand, rubber and other components to produce the natural turf-like effect.

"Although this machine is custom-made for today's artificial turf surfaces, it's another example of how versatile the tufting machine has become," Lewis said.

"After all the years I was in the business, seeing the pile fabrics being made on tufting machines, it doesn't surprise me what they can do with it. All of what is made that we call pile fabrics – bedspreads, throw rugs, carpet, turf – it just doesn't surprise me any more, although it amazes me sometimes. It's basically the same process, no matter what you're producing – just different applications for the same process."

Sticking With the Basics

Tufted turf is just one of many possibilities for the tufting process. In a recent episode of a popular TV home improvements-makeover show, one wall of a bedroom was decorated with carpet. It could be that at some point in the future, carpet will fulfill the prediction by a prominent designer that it would become common as wall covering as well as floor covering.

Whether this comes to pass remains to be seen, but as Charlie expressed it, CMC's immediate plans are to stick with the basics while sustaining its passion for improvement.

"We see ourselves doing what we have been doing, only doing it better, constantly learning and creating possibilities. This is a daily reality at CMC. Electronic technology is certainly helping us to do things we could not do before, and we are learning how to use and apply this technology in different ways.

"Our benchmark is still woven carpet – and we are getting closer and closer to being able to do virtually anything that can be accomplished through the weaving process.

"At a recent FloorTek show in Dalton, where the latest in flooring and tufting technology was displayed, the representative of a European loom manufacturer kept coming by our booth where we had samples of our Virtual Weave cut pile. He said if someone had asked him if that could be done on a tufting machine, he would have emphatically said no, but he could see that it was possible after all. All he could do was marvel.

"So as we pursue new styling developments, we always benchmark woven carpet to see what it offers that we can't do yet."

Just as Lewis and Roy were the pacesetters for innovation in years past, today at CMC the plant manager Allen Neely, along with Bill Christman and Wilton Hall, serve as leaders in exploring new horizons. However, the innovation responsibilities have been extended considerably. "We have a broader spectrum of people involved today," Charlie said. "We view everyone as being responsible for product development. It's a collaborative, team effort.

"Every customer we have is involved, too, offering their ideas and suggesting what they would like our machines to be able to do," Lewie pointed out. "We have a dedicated research and development department at our Hixson facility, as well as Sonny and Heath Jackson at our Jackson Plant."

"I am continually amazed at what extremely creative people can do," he said. "We give them a better technology, and it is incredible what they are able to accomplish with it. They start to apply it in all of the different areas where floor covering is being furnished.

"Of course, we have also learned that some technology – no matter how impressed we are with it – may not generate interest in the marketplace. It all depends on the fashion trends at the time. The good news is, just as with clothing, what goes around comes around. What's not popular today may become the rage tomorrow with a new twist here or there."

Even though there has been some indication that consumer tastes are shifting somewhat from carpet, Lewie says there is no apprehension that demand for their products will sharply decline or even disappear. "The new stuff has settled in and we have succeeded in retaining our floor space. There are new opportunities with other products, such as rugs, as well as manufacturing other machines to produce different kinds of floor coverings.

"I remember 20 years ago an industry consultant, Reg Burnett, stated at a Dalton civic gathering that the tufting machine would become a thing of the past, being replaced by the fusion-bonding method of manufacturing carpet. But today the fusion-bonding machine is virtually obsolete and we keep reinventing the tufting process ourselves, developing it into a better, more productive and more innovative method. That keeps us in business.

"Artificial turf is a growing market all over the world, which creates a new opportunity for our machines, and there are other specialty areas, such as geotextiles – tufting liners for landfills. But ultimately, 99.9% of what we do will remain in the area of floor coverings."

Riding Out Tough Economic Waters

For most industries, 2009 was the worst year economically in recent memory, due to the worldwide recession – a year that will not soon be forgotten. The floor covering industry was no exception. With carpet manufacturers seeing sales volumes fall drastically, CMC's production and sales underwent a similar decline.

"In 2009, our business was down about 50 percent compared to 2008," Charlie said, "and the last quarter of 2008 was not very good. We had to temporarily lay off a lot of our people, and we had to cut back work hours and compensation for everyone that remained.

"Thankfully, we were able to get through the major part of the recession.

We've been able to rehire most of the people we laid off last year, along with restoring the salaries for staff who stayed on. Part of the challenge of late has been working with vendors and suppliers to get our business ramped back up to meet the volume of new business that has come in the past few months."

While the domestic flooring market has experienced a slow recovery, CMC's emphasis on international trade has paid off. "We're seeing a substantial return on strategic plans we have implemented over the past several years, and we feel our market share globally is much higher today because of our direct sales effort, shifting from working through independent agents in different parts of the world," Charlie noted.

"Even in the toughest of times, we remained confident that we were right where God wanted us to be and that He would continue to sustain the work at CMC. We give glory to God for enabling us to weather this storm – the worst we've seen to this point. He always has been, and continues to be, our Provider."

Lewie agreed. "We have a lot of very capable people here at CMC, but even at that we're not that smart to be able to rebound as we have. It's all in God's hands, and we're very grateful that He continues to work in our company – and through it."

Ongoing Commitment to Quality

Charlie said marketing remains a key for CMC's future success, although not in terms of acquiring new customers. Rather, their marketing focus is on sustained and enhanced quality and service. "Early on we knew that we could sell the president of a carpet mill one machine, but it was the people on the shop floor, the ones who actually work on the machines day after day – they were the people who would ensure that we received the orders for the second, 22nd, or 202nd machines. So our challenge was – and continues to be – to build the machines as simply, easy to understand, as easy to work on and as trouble-free as possible.

"And because our machines are so well-built, the key to our continued business is creating obsolescence with new technology. Since carpet is a styling business, in the future we must continue to give our customers, the leaders in the industry, what they are clamoring for. In many cases, this means new styling developments. The heartbeat of this business is new product development – it helps to build machines that are faster and more efficient, but the heartbeat is being able to develop new products."

When CMC started, there were more than 200 independent carpet mills in the United States alone, but through the consolidation of ownership, that number is greatly reduced. Today CMC's customers throughout the world total just over 200 – a finite market.

"We could spend millions of dollars on advertising, but it would not get us one more customer," Lewie said. "In reality, our market is extremely limited; we already know everyone that is out there and they know us. Advertising is important to companies like Coca-Cola, Little Debbie and other consumer products. But for us, our money is best spent on development ideas and constantly re-inventing what we do and how we do it."

Charlie said he was particularly pleased by an example of how CMC's commitment to lean manufacturing is taking hold throughout the organization.

"During the recent economic downturn, we seized the opportunity to work toward improving our processes and documenting attention to detail. We believed that in spite of challenges confronting the entire industry, we could come out of this an even better company.

"Recently some of our guys on the floor, not supervisors, were going through lean training certification, which is designed for them to gain a stronger grasp of the concepts. One of the men stood before about 25 other people and, using a PowerPoint presentation, discussed ways for reducing waste of motion, time and materials – and showed tangible results.

"It had to be one of the best days of my life here at CMC. It showed that our commitment to lean has taken on a life of its own, including our workers on the shop floor. It's no longer Jim Joyner's idea, or that of the executives or our management team. The entire CMC family is catching on to its importance."

Operating According to Foundational Values

From the start, the accomplishments of Lewis and Roy Card were the result of their diligent application of God-given talents and imagination. Their innate traits of determination, creativity, hard work and the pursuit of excellence drove them to achieve the unprecedented and unimaginable. Today CMC, the latest link in the tufting chain that started with Cobble Brothers Machinery, is expressly dedicated to honoring the God who provided their inspiration.

As Brad Card noted, "In our Mission Statement, it states that our work is 'All to the Glory of God.' This is a continual reminder that our goal is not just to please our customers, but also our Creator. Do we always succeed? No. But this is why it also states in our Values statement, 'If you don't believe we are living up to these values, please let us know.' But to please our customers – and God – that is our goal.

"We realize our business ultimately has come from the Lord and know we have a high standard to strive for – we understand God expects that of us. Because of this, we place a high value on people – not only our employees, but also our customers and everyone we do business with. People are the most important thing to CMC.

"We would not exist if it were not for our customers and the employees who do the work here. More than that, we seek to view people in terms of the value they have as individuals."

Legacies That Endure

Lewie Card acknowledged a simple yet significant ritual his father always has observed upon entering the CMC lobby. "When Dad visits, he never fails to touch the bust of Joe Cobble that we have displayed in the lobby (along with busts of Lewis and Roy). This is not a superstitious or mystical gesture, but just his way of always conveying his grateful heart. Dad reminds us frequently that all of what CMC has accomplished – not to mention what has transpired in the tufting industry overall – could not have been possible if it had not been for Joe Cobble. He provided the setting, an environment for exploring the horizons of what could be done in tufting."

The busts of Lewis and Roy, literally sharing the lobby spotlight with Joe's, stand in tribute to their own legacies. "We have to thank Dad and Roy for the opportunity they gave to us by paving the way, passing on to us such a rich heritage of diligence and inventiveness," Lewie said. "They have been a true blessing to me, and what we have done would not have been possible without them. We're thankful for being in this position today."

Statistics have shown that only about one-third of family-owned businesses pass successfully to the second generation, and the success rate of surviving into the third generation falls much lower. The fact that CMC is now being led by the third generation of the Card family in the industry and already has members of the family's fourth generation vitally involved is not taken lightly.

"What a blessing it has been to be in business with family," Lewie asserted, "sharing such interesting and challenging times and opportunities together."

"Families in business can either be the best thing in the world, or the worst thing in the world," Charlie added. "Fortunately for us, it has been the best thing in the world."

What has been the secret? There was never any magical secret, he commented. "Somehow we have always managed to be of like mind and one accord. We have not always agreed in every circumstance, but we have agreed to pull together. Families pull apart when there are conflicts of egos, ethics, greed, or differences in mission, overriding goals and values.

"Even in tough times we have been able to remain pretty much on the same page. Rather than weaken, divide or even destroy the company, at CMC our family relationships have helped and served to strengthen our company and our mission."

"Our hope is to pass this legacy on," Lewie said, "honoring our predecessors and seeing the examples and values continue that they so graciously provided for us. We want the next generation to understand what got us here, being able to appreciate and honor the legacy that has been handed down to us."

A Carpet of Many Colors

In the main lobby of CMC, one of the first things visitors notice is the carpet that displays a variety of designs and colors – deep red, tan, beige, checked (black, tan, gray and beige) and taupe – creatively installed together. This reflects some of the various types of carpet that can be realized today through tufting.

As innovations continue, that carpet one day may give way to another that more accurately represents the advancements in tufting technology at that time. What those changes might be, no one can state for certain. But then, no one could have predicted how a modified industrial sewing machine slowly making tufted bedspreads one at a time would become the precursor of huge, multi-million dollar machines with thousands of needles that produce miles and miles of luxurious, colorful, textured carpet.

It has been a remarkable journey from the rustic home of Catherine Evans to the shops of Joe and Albert Cobble to the companies founded by Lewis and Roy Card to state of the art facilities of CMC. What the next destinations will be one can only guess, but chances are they will be, to repeat the word frequently used in this chapter: *Amazing.*

Timeline

1937	Cobble Bros. established by Joe & Albert Cobble, 315 East Main Street, Chattanooga, Tennessee
1938	Chenille bedspread equipment and single needle tufting machine introduced
1939	Joe Cobble hires Lewis Card, Sr.
1940	Cobble Bros. moves to 315 West Main Street First 9-foot bedspread yardage machine introduced
1943	Lewis Card, Sr. promoted to general manager of Cobble Bros.
1944	Cobble Bros. manufactured 90mm gun parts for Wheland Foundry for World War II
1947	Chenille bedspread equipment [table models and yardage machines] Table model multi-needle rug machines in use by 1947
1950	Lewis Card, Sr. hires Roy Card
1951	Cobble begins producing tufted carpet machines Lewis Card, Sr. implements yarn feed for tufting machines
1952	Four-roll pattern attachment introduced

1953	Roy Card moves to Sample/Experimental Dept. at Cobble Bros.
1955	Cobble Bros. moves to Riverside Drive Scroll attachment introduced
1956	Drum Pattern-control attachment introduced drum with fingers that were electrical switches
1957	Cobble Bros. purchases Super Tufter from Albert Cobble
1958	Roy Card promoted to manager of Service Dept. and Sample/Experimental Dept.
1959	Cobble acquires British Tufting Machine and started Cobble England
1960	Singer buys Cobble Bros., changes name to Singer-Cobble, and sells Southern Machine back to Joe and Albert Cobble
1962	Roy Card invents Loop Cut Mechanism
1963	Lewis Card, Sr. leaves Singer-Cobble
1964	Roy Card Co. and Lewis Card & Co. were founded Drum pattern-control attachment revised to use up to 120 photoelectric cells
1965	Lewis Card, Sr. and Roy Card merge and form Card & Company, Inc.
1966	Joe Cobble passes away
1969	Card & Co. and Southern Machine combine to form Tuftco public offering
1971	Lewis Card, Jr. joins Card & Co.
1972	Charles Monroe joins Card & Co.
1974	Hydrashift (hydraulic shifting) at Card & Co. introduced
1977	Lewis Card, Sr. and Roy Card sell Tuftco; they continue to work there

1980	Lewis Card, Sr. and Roy Card leave Tuftco at the end of 1980
1981	Charles Monroe Company (CMC) founded by Charles Monroe and Lewis Card, Jr. (April)
1983	Charles Monroe Company changes name to Card-Monroe Corp. Lewis Card, Sr. and Roy Card join Card-Monroe Corp. (CMC) CMC Advanced Cutting System introduced
1985	SuperSpeed Machine introduced by CMC
1989	Command Performance: computer-based, servo-driven control system introduced by CMC to automate the tufting machine
1995	Servo Scroll introduced by CMC
2003	Infinity Attachment introduced by CMC
2008	ColorPoint introduced by CMC
2010	Lewis Card, Sr. and Roy Card inducted into the University of Tennessee-Chattanooga (UTC) Entrepreneurship Hall of Fame

Basic Tufting Terminology

Gauge: The density or positioning of yarns, as determined by the distance between two adjacent needle points. Usually stated in fractions of an inch, a 1/8-gauge has needles each 1/8-inch apart.

Stitches per inch: This is the number of yarn tufts per inch of a single tuft row in tufted carpet. The carpet face weight and density are controlled by the number of stitches per inch.

Primary backing: This is material supplied from a roll in front of the tufting machine. Spiked rolls on the front and back of the machine feed the backing through the machine.

Looper (or hook): This device, shaped like an inverted hockey stick, catches yarn from the needles and holds it to form loops. To create cut pile, a looper and knife combination is used to cut the loops. For cut-loop combinations, a special looper and conventional cutting knife are used in concert.

Pile height: This is typically measured from the surface of the primary backing to the top of tufted yarn.

Level loop pile: All loops on the carpet surface are the same pile height.

Cut pile: Each tuft in the carpet is cut by a knife as the yarn is drawn through the primary backing by the looper.

Level cut loop pile: The carpet surface is a combination of cut and looped yarns all at the same pile height.

High cut-low loop pile: The surface pile of the carpet is a combination of cut and looped yarns.

High-low patterned loop: Surface patterns created by varying the pile heights of individual loops of yarn.

Graphics pattern: This applies to the variety of designs created by stepping, or zigzag, moving needle bars and individually specific threading sequences, making possible a wide variety of patterns.

Creel: Racks of yarn cones, holding yarn that is fed into the tufting machine through guide tubes.

Types Of Tufting Construction

Loop Pile

Cut Pile

Level Cut Loop

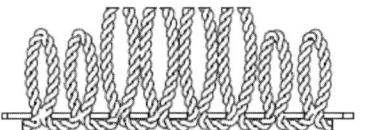

High Cut Low Loop

High Low Patterned Loop

Lewis Card, Sr. (left) and his brother, Roy Card, were inducted into the
University of Tennessee-Chattanooga's School of Business Entrepreneurship
Hall of Fame, in April of 2010. (Photo by Barry Aslinger)

Bibliography

Beasley, Marion Max, *Memoirs*

Deaton, Thomas M., *Bedspreads to Broadloom: The Story of the Tufted Carpet Industry,* Tapestry Press, Acton, Mass., 1993.

Longwith, John, *Bound and Determined: The Dixie Group, 1920-2004,* Chattanooga, Tenn., 2005.

Patton, Randall L., *Carpet Capital: The Rise of a New South Industry,* Athens, Ga., University of Georgia Press, 1999.

Patton, Randall L., *Shaw Industries: A History,* Athens, Ga., University of Georgia Press, 2002.

The Carpet and Rug Institute, *CRI: The Carpet Primer,* Dalton, Ga., 1995.

The Carpet and Rug Institute, *Carpet Makes the World Better,* Dalton, Ga.

Webb, James, *Born Fighting: How the Scots-Irish Shaped America,* New York: Broadway Books, 2004.

About the Author

Robert J. Tamasy is vice president of communications for Leaders Legacy, Inc., based in Atlanta, Georgia, a non-profit organization that serves business and professional leaders through leadership development, executive coaching and mentoring. A former community and suburban newspaper editor, Bob has written, co-authored and edited more than a dozen books and hundreds of magazine articles, specializing in business and workplace topics. He and his wife, Sally, reside in Chattanooga, Tennessee, and together have five children and eight grandchildren.